W9-BWG-429

The Forest Unseen

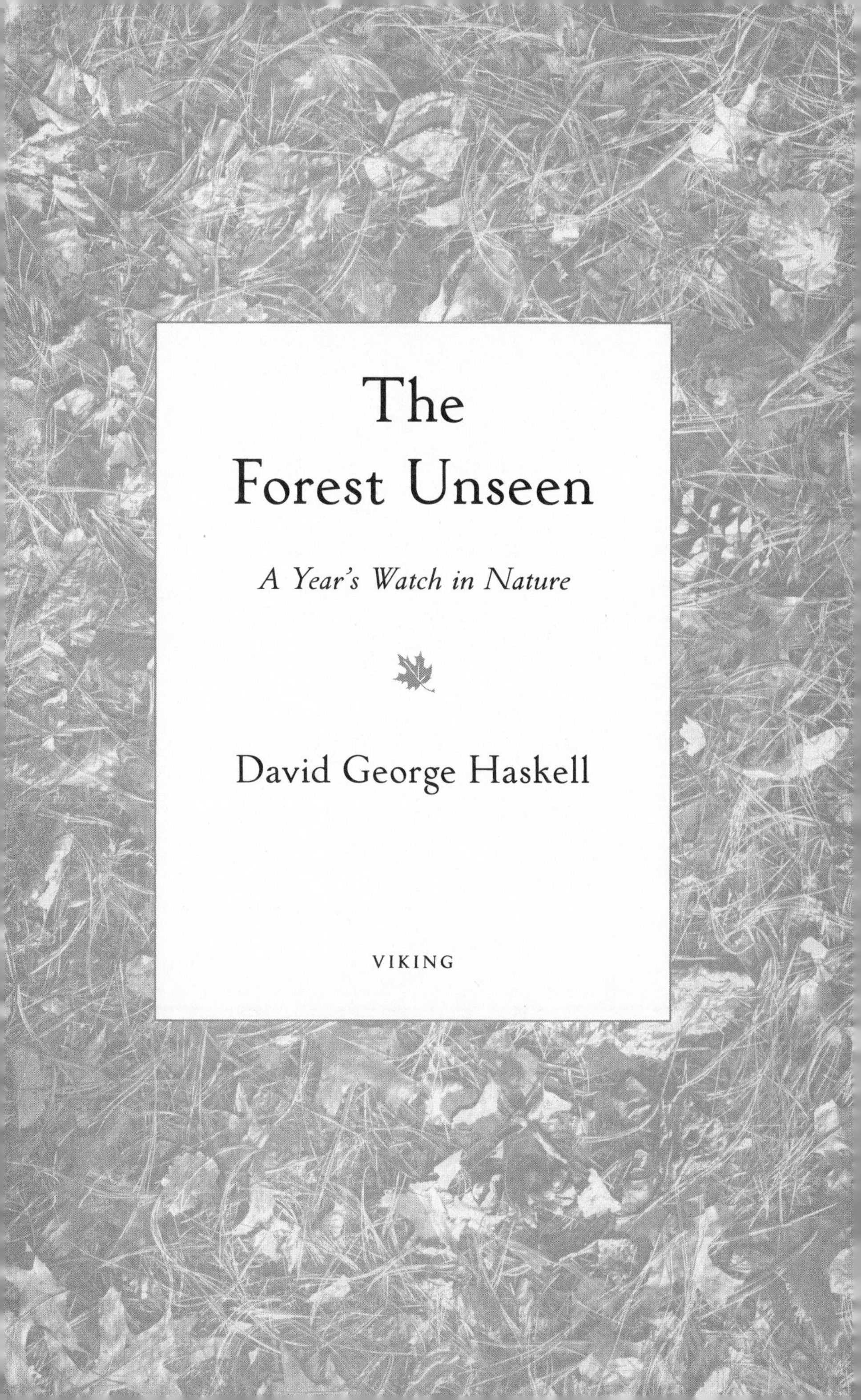

The Forest Unseen

A Year's Watch in Nature

David George Haskell

VIKING

VIKING
Published by the Penguin Group
Penguin Group (USA) Inc., 375 Hudson Street, New York, New York 10014, U.S.A.
Penguin Group (Canada), 90 Eglinton Avenue East, Suite 700, Toronto, Ontario, Canada M4P 2Y3
(a division of Pearson Penguin Canada Inc.)
Penguin Books Ltd, 80 Strand, London WC2R 0RL, England
Penguin Ireland, 25 St. Stephen's Green, Dublin 2, Ireland (a division of Penguin Books Ltd)
Penguin Books Australia Ltd, 250 Camberwell Road, Camberwell, Victoria 3124, Australia
(a division of Pearson Australia Group Pty Ltd)
Penguin Books India Pvt Ltd, 11 Community Centre, Panchsheel Park, New Delhi – 110 017, India
Penguin Group (NZ), 67 Apollo Drive, Rosedale, Auckland 0632, New Zealand
(a division of Pearson New Zealand Ltd)
Penguin Books (South Africa) (Pty) Ltd, 24 Sturdee Avenue, Rosebank,
Johannesburg 2196, South Africa

Penguin Books Ltd, Registered Offices: 80 Strand, London WC2R 0RL, England

First published in 2012 by Viking Penguin, a member of Penguin Group (USA) Inc.

5 7 9 10 8 6

LIBRARY OF CONGRESS CATALOGING IN PUBLICATION DATA
Haskell, David George.
The forest unseen : a year's watch in nature / David George Haskell.
p. cm.
Includes bibliographical references and index.
ISBN 978-0-670-02337-0
1. Old growth forest ecology—Tennessee. 2. Old growth forests—Tennessee. 3. Natural history—Tennessee. 4. Seasons—Tennessee. 5. Nature observation—Tennessee. 6. Haskell, David George. 7. Philosophy of nature. I. Title.
QH105.T2H37 2012
577.309768—dc23 2011037552

Printed in the United States of America
Designed by Nancy Resnick

For Sarah

Contents

Preface

Two Tibetan monks lean over a table, cradling brass funnels in their hands. Colored sand spills from the tips of the funnels onto the table. Each fine stream adds another line to the growing mandala. The monks work from the center of the circular pattern, following chalk lines that define the fundamental shapes, then filling in hundreds of details from memory.

A lotus flower, symbol of Buddha, lies at the center and is enclosed by an ornate palace. The four gates of the palace open out to concentric rings of symbols and color, representing steps on the path to enlightenment. The mandala will take several days to complete, then it will be swept up and its jumbled sands cast into running water. The mandala has significance at many levels: the concentration required for its creation, the balance between complexity and coherence, the symbols embedded in its design, and its impermanence. None of these qualities, however, define the ultimate purpose of the mandala's construction. The mandala is a re-creation of the path of life, the cosmos, and the enlightenment of Buddha. The whole universe is seen through this small circle of sand.

A group of North American undergraduates jostle behind a rope nearby, extending their necks like herons as they watch the mandala's birth. They are uncharacteristically quiet, perhaps caught up in the work or stilled by the otherness of the monks' lives. The students are visiting the mandala at the beginning of their first laboratory class in ecology. The class will continue in a nearby forest, where the students will create their own mandala by throwing a hoop onto the ground. They will study their circle of land for the rest of the afternoon, observing the workings of the forest community. One translation of the Sanskrit *mandala* is "community," so the monks and the students are engaged in the same work: contemplating a mandala and refining their minds. The parallel runs deeper than this congruence of language and symbolism. I believe that the forest's ecological stories are all present in a mandala-sized area. Indeed, the truth of the forest may be more clearly and vividly revealed by the contemplation of a small area than it could be by donning ten-league boots, covering a continent but uncovering little.

The search for the universal within the infinitesimally small is a quiet theme playing through most cultures. The Tibetan mandala is our guiding metaphor, but we also find context for this work in Western culture. Blake's poem "Auguries of Innocence" raises the stakes by shrinking the mandala to a speck of earth or a flower: "To see a World in a Grain of Sand / And a Heaven in a Wild Flower." Blake's desire builds on the tradition of Western mysticism most notably demonstrated by the Christian contemplatives. For Saint John of the Cross, Saint Francis of Assisi, or Lady Julian of Norwich, a dungeon, a cave, or a tiny hazelnut could all serve as lenses through which to experience the ultimate reality.

This book is a biologist's response to the challenge of the Tibetan mandala, of Blake's poems, of Lady Julian's hazelnut. Can the whole forest be seen through a small contemplative window of leaves, rocks, and water? I have tried to find the answer to this question, or the start of an answer, in a mandala made of old-growth forest in the hills of Tennessee. The forest mandala is a circle a little over a meter across, the same size as the mandala that was created and swept away by the monks. I chose the mandala's location by walking haphazardly through the forest and stopping when I found a suitable rock on which to sit.

The area in front of the rock became the mandala, a place that I had never seen before, its promise mostly hidden by winter's austere garb.

The mandala sits on a forested slope in southeastern Tennessee. One hundred meters upslope, a high sandstone bluff marks the western edge of the Cumberland Plateau. The ground falls away from this bluff in steps, alternating level benches with sharp inclines, descending one thousand feet in elevation to the valley floor. The mandala nestles between boulders on the highest bench. The slope is entirely forested with a diverse collection of mature deciduous trees: oaks, maples, basswoods, hickories, tuliptrees, and a dozen more species. The forest floor is ankle-twistingly strewn with jumbled rock from the eroding bluff, and in many places there is no even ground, just heaved, fissured stone overlain with leaf mulch.

The steep, challenging terrain has protected the forest. At the bottom of the mountain, the fertile, level soil on the valley floor is relatively free of rocky encumbrances and has been cleared for pasture and crops, first by Native Americans, then by settlers from the Old World. A few homesteaders tried to farm the mountainside in the late nineteenth and early twentieth centuries, a task that was as hard as it was unproductive. Moonshine stills gave these subsistence farmers extra income, and this mountainside got its name, Shakerag Hollow, from the way townspeople would summon the distillers by waving a rag that was then left with some money. A jar of strong liquor would take the place of the money some hours later. The forest has now reclaimed the small agricultural openings and still sites, although the locations of the old clearings are marked by rock heaps, old pipes, rusted washtubs, and daffodil patches. Much of the rest of the forest was logged for lumber and fuel, especially at the turn of the twentieth century. A few small pockets of forest were left untouched, shielded by inaccessibility, luck, or the whims of landowners. The mandala sits in one such patch, a dozen or so acres of old-growth forest embedded in thousands of acres of forest that, although it has been cut in the past, is now mature enough to sustain much of the rich ecology and biological diversity that characterize Tennessee's mountain forests.

Old-growth forests are messy. Within a stone's throw of the mandala, I see half a dozen large fallen trees in various stages of decomposition. The rotting logs are the food for thousands of species of animals, fungi, and microbes. Downed trees leave gaps in the forest canopy, creating the second characteristic of old-growth forests, a mosaic of tree ages, with groups of young trees growing next to thick-trunked elders. A pignut hickory with a trunk a meter wide at its base grows just to the west of the mandala, right next to a crowd of maple saplings in a gap left by a massive fallen hickory. The rock on which I sit is backed by a middle-aged sugar maple, its trunk as wide as my torso. This forest has trees of all ages, a sign of the historical continuity of the plant community.

I sit next to the mandala on a flat slab of sandstone. My rules at the mandala are simple: visit often, watching a year circle past; be quiet, keep disturbance to a minimum; no killing, no removal of creatures, no digging in or crawling over the mandala. The occasional thoughtful touch is enough. I have no set schedule for visits, but I watch here many times each week. This book relates the events in the mandala as they happen.

The
Forest Unseen

January 1st—Partnerships

The New Year starts with a thaw, and the fat, wet smell of the woods fills my nose. Moisture has plumped the mat of fallen leaves that covers the forest floor, and the air is suffused with succulent leafy aromas. I leave the foot trail that winds down the forest slope and scramble around a house-sized piece of mossy, eroded rock. Across a shallow bowl on the mountainside I see my landmark: a long boulder, cresting out of the leaf litter like a small whale. This block of sandstone defines one edge of the mandala.

It takes me just a few minutes to traverse the rocky scree and reach the boulder. I step past a large hickory tree, resting my hand on its gray strips of bark, and the mandala is at my feet. I circle to the opposite side and take my seat on a flat rock. After pausing to inhale the rich air, I settle in to watch.

The leaf litter is mottled with browns. A few bare spicebush stems and a young ash stand waist-high at the mandala's center. The muted, leathery colors of these decaying leaves and dormant plants are eclipsed by the glow coming from the rocks that frame the mandala. These stones are tumbled remnants of the eroding bluff, smoothed into irregular, lumpy forms by thousands of years of erosion. The rocks range in size from woodchuck to elephant; most are about as big as a curled-up human. Their radiance comes not from stone but from mantles of lichen that blush emerald, jade, and pearl in the humid air.

The lichens' growth forms mountains in miniature, sandstone crags with variegated patches of moisture and sunlight. The highest ridges on the boulders are spattered with tough-skinned gray flakes. Dark canyons between rocks have a purple sheen. Turquoise glistens on vertical walls, and concentric circles of lime flow down gentle slopes. All the lichens' hues are paint-stroke fresh. This vibrancy contrasts with the winter-weighed lethargy of the rest of the forest; even the mosses are muted and frost-bleached.

Supple physiology allows lichens to shine with life when most other creatures are locked down for the winter. Lichens master the cold months through the paradox of surrender. They burn no fuel in quest of warmth, instead letting the pace of their lives rise and fall with the thermometer. Lichens don't cling to water as plants and animals do. A lichen body swells on damp days, then puckers as the air dries. Plants shrink back from the chill, packing up their cells until spring gradually coaxes them out. Lichen cells are light sleepers. When winter eases for a day, lichens float easily back to life.

This approach to life has been independently discovered by others. In the fourth century BCE, the Chinese Taoist philosopher Zhuangzi wrote of an old man tossed in the tumult at the base of a tall waterfall. Terrified onlookers rushed to his aid, but the man emerged unharmed and calm. When asked how he could survive this ordeal, he replied, "acquiescence . . . I accommodate myself to the water, not the water to me." Lichens found this wisdom four hundred million years before the Taoists. The true masters of victory through submission in Zhuangzi's allegory were the lichens clinging to the rock walls around the waterfall.

The quietude and outer simplicity of the lichens hides the complexity of their inner lives. Lichens are amalgams of two creatures: a fungus and either an alga or a bacterium. The fungus spreads the strands of its body over the ground and provides a welcoming bed. The alga or bacterium nestles inside these strands and uses the sun's energy to assemble sugar and other nutritious molecules. As in any marriage, both partners are changed by their union. The fungus body spreads out, turning itself into a structure similar to a tree leaf: a protective upper crust, a layer for the light-capturing algae, and tiny pores for breathing. The algal partner loses its cell wall, surrenders protection to the fungus, and gives up sexual activities in favor of faster but less genetically exciting self-cloning. Lichenous fungi can be grown in the lab without their partners, but these widows are malformed and sickly. Similarly, algae and bacteria from lichens can generally survive without their fungal partners, but only in a restricted range of habitats. By stripping off the bonds of individuality the lichens have produced a world-conquering union. They cover nearly ten percent of the land's surface, especially in the treeless far north, where winter reigns for most of the year. Even here, in a tree-filled mandala in Tennessee, every rock, trunk, and twig is crusted with lichen.

Some biologists claim that the fungi are exploiters, ensnaring their algal victims. This interpretation fails to see that the lichen partners have ceased to be individuals, surrendering the possibility of drawing a line between oppressor and oppressed. Like a farmer tending her apple trees and her field of corn, a lichen is a melding of lives. Once individuality dissolves, the scorecard of victors and victims makes little sense. Is corn oppressed? Does the farmer's dependence on corn make her a victim? These questions are premised on a separation that does not exist. The heartbeat of humans and the flowering of domesticated plants are one life. "Alone" is not an option: the farmer's physiology is sculpted by a dependence on plants for food that dates back hundreds of millions of years to the first wormlike animals. Domesticated plants have experienced only ten thousand years of life with humans, but they too have shed their independence. Lichens add physical intimacy to this interdependence, fusing their bodies and intertwining the membranes of their cells, like cornstalks fused with the farmer, bound by evolution's hand.

The diversity of color in the lichens on the mandala reflects the many types of algae, bacteria, and fungi involved in the lichens' union.

Blue or purple lichens contain blue-green bacteria, the cyanobacteria. Green lichens contain algae. Fungi mix in their own colors by secreting yellow or silver sunscreen pigments. Bacteria, algae, fungi: three venerable trunks of the tree of life twining their pigmented stems.

The algae's verdure reflects an older union. Jewels of pigment deep inside algal cells soak up the sun's energy. Through a cascade of chemistry this energy is transmuted into the bonds that join air molecules into sugar and other foods. This sugar powers both the algal cell and its fungal bedfellow. The sun-catching pigments are kept in tiny jewel boxes, chloroplasts, each of which is enclosed in a membrane and comes with its own genetic material. The bottle-green chloroplasts are descendants of bacteria that took up residence inside algal cells one and a half billion years ago. The bacterial tenants gave up their tough outer coats, their sexuality, and their independence, just as algal cells do when they unite with fungi to make lichens. Chloroplasts are not the only bacteria living inside other creatures. All plant, animal, and fungal cells are inhabited by torpedo-shaped mitochondria that function as miniature powerhouses, burning the cells' food to release energy. These mitochondria were also once free-living bacteria and have, like the chloroplasts, given up sex and freedom in favor of partnership.

And life's chemical whorl, DNA, bears the marks of yet more ancient union. Our bacterial ancestors shuffled and swapped their genes among species, blending genetic instructions like cooks copying from one another's recipe cards. Occasionally two chefs would agree to a wholesale merger, and two species fused into one. The DNA of modern organisms, including our own, retains traces of such mergers. Although our genes function as one unit, they come with two or more subtly different writing styles, vestiges of the different species that united billions of years ago. The "tree" of life is a poor metaphor. The deepest parts of our genealogies resemble networks or deltas, with much interweaving and cross flow.

We are Russian dolls, our lives made possible by other lives within us. But whereas dolls can be taken apart, our cellular and genetic helpers cannot be separated from us, nor we from them. We are lichens on a grand scale.

Union. Fusion. The mandala's inhabitants are plaited into winning partnerships. But cooperation is not the only relationship in the forest. Piracy and exploitation are here also. A reminder of these more painful associations lies coiled on the leaf litter at the center of the mandala, enclosed by the lichen-coated rocks.

The reminder unwrapped itself slowly, held back by the torpidity of my powers of observation. My attention was first drawn by two amber ants bustling across the wet leaf litter. I watched their scurrying for half an hour before I noticed the ants' particular interest in a coiled strand nestled in the litter. The strand was about as long as my hand and was the same rain-soaked brown as the hickory leaf on which it lay. At first I dismissed the curl as an old vine tendril or leaf stem. But as my eyes were about to move on to more stimulating things, an ant paddled the tendril with her antennae, and the coil straightened and lurched. My mind started into recognition: a horsehair worm. A strange creature, with a taste for exploitation.

The worm's twisting gave away its identity. Horsehair worms are pressurized from within, and the tug of muscles against this inflated body makes the worm jerk and writhe like no other animal. The worm has no need of complicated or graceful movement, for at this stage of its life it has only two tasks left, to squirm toward a mate and then to lay eggs. Nor did the worm have need of sophisticated motion in its previous life stage, when it lay balled inside the body of a cricket. The cricket did the worm's walking and feeding. The horsehair worm lived as an internal brigand, robbing then killing the cricket.

The worm's life cycle began when it hatched from an egg laid in a puddle or stream. The microscopic larva crawled over the streambed until it was eaten by a snail or small insect. Once inside its new home, the larva wrapped itself in a protective coat, formed a cyst, then waited.

The lives of most larvae are cut short at this point, as cysts, never completing the rest of the life cycle. The worm in the mandala was one of the few that make it to the next stage. Its host crawled onto land, died, and was chewed on by an omnivorous cricket. This is such an improbable sequence of events that the horsehair worm's life cycle requires parent worms to lay tens of millions of eggs; on average, only one or two of this multitude of young will survive to adulthood. Once inside the cricket, the spiny-headed larval pirate bored through the gut wall and took up residence in the hold, where it grew from a comma-sized larva to a worm the length of my hand, coiling upon itself to fit within the cricket. When the worm could grow no more, it released chemicals that took over the cricket's brain. The chemicals turned the water-fearing cricket into a suicidal diver seeking puddles or streams. As soon as the cricket hit water, the horsehair worm tensed its strong muscles, ripping through the cricket's body wall, and twisted free, leaving the plundered vessel to sink and die.

Once free, horsehair worms have a keen appetite for company, and they mate in untidy skeins of tens or hundreds of worms. This habit has given them a second name, Gordian worms, after the eighth-century legend of King Gordius's monstrously complex knot. Whoever could unbind the knot would succeed the king, but all would-be rulers failed. It took another pirate, Alexander the Great, to loose the knot. Like the worms, he cheated his hosts, slashing the knot with his sword and claiming the country's crown.

After the Gordian mating tangle is sated, the worms unwrap and crawl away. They lodge their eggs in soggy pond margins and damp forest floors. Once hatched, the worm larvae will pick up the Alexandrian plunderer's spirit, first infecting a snail, then emerging to rob a cricket.

The horsehair worm's relationship with its hosts is entirely exploitative. Its victims receive no hidden benefit or compensation for their suffering. But even this parasitic worm is sustained by an interior crowd of mitochondria. Piracy is powered by collaboration.

. . .

Taoist union. Farmer's dependence. Alexandrian pillage. Relationships in the mandala come in multifarious, blended hues. The line between bandit and honest citizen is not as easily drawn as it first seems. Indeed, evolution has drawn no line. All life melds plunder and solidarity. Parasitic brigands are nourished by cooperative mitochondria within. Algae suffuse emerald from ancient bacteria and surrender inside gray fungal walls. Even the chemical ground of life, DNA, is a maypole of color, a Gordian knot of relationships.

January 17th—Kepler's Gift

Ankle-deep snow has smoothed the forest's fractured, uneven surface into a gentle swell and trough. This covering disguises deep cracks between rocks and makes walking treacherous. I move slowly, bracing myself against tree trunks as I slide and clamber to the mandala. I brush the snow from my rock, then sit, huddling in my coat. Loud cracks, like gunfire, echo down the valley every ten minutes or so, the sound coming from snapping fibers in ice-stiffened branches of the bare, gray trees. The temperature has dropped to ten below, not a hard freeze but the first real cold of the year and enough to stress the trees' wood.

The sun emerges, and snow transforms from a soft layer of white into thousands of sharp, bright points of light. I hook a fingertip of this glittering jumble from the mandala's surface. Seen closely, the snow is a tangle of mirrored stars, each one flashing as its surface aligns with the sun and my eye. The sunlight catches the minute ornamentation of each flake, revealing perfectly symmetrical arms, needles, and hexagons. Hundreds of these exquisite ice flakes crowd onto one fingertip.

How is such beauty born?

In 1611, Johannes Kepler took time away from elucidating the motions of the planets to meditate on the snowflake. He was particularly intrigued by the regularity of snowflakes' six sides: "There must be some definite cause why, whenever snow begins to fall, its initial formations invariably display the shape of a 6-sided starlet." Kepler searched for an answer in the rules of mathematics and the patterns of natural history. He noted that the honeybee and pomegranate array their combs and seeds in hexagons, perhaps reflecting geometrical efficiency. But water vapor is not squeezed into a rind like pomegranate seeds, nor is it built up by the work of insects, so Kepler believed that these living examples could not reveal the cause of the snowflakes' architecture. Flowers and many minerals don't conform to the six-sided rule, further frustrating Kepler's search. Triangles, squares, and pentagons can also be stacked into neat geometrical patterns, eliminating pure geometry from the list of possibilities.

Kepler wrote that snowflakes are showing us the spirit of the earth and God, the "formative soul" that inhabits all being. But this medieval solution didn't satisfy him. He sought a material explanation, not a finger pointing toward mystery. Kepler ended his essay in frustration, unable to peer beyond the door of the icy palace of knowledge.

His frustration could have been eased had he taken seriously the concept of the atom, an idea that originated with the classical Greek philosophers but had fallen out of favor with Kepler and most early seventeenth-century scientists. Yet the atoms' two-thousand-year exile was coming to a close, and by the end of the seventeenth century atoms became fashionable again, balls and sticks dancing triumphant across textbooks and chalkboards. Now, we seek out atoms by blasting ice with X-rays, using the pattern of the rays that emerge to discover a world one million billion times smaller than the normal scale of human life. We find jagged lines of oxygen atoms, each atom tethered to two restless hydrogen atoms, electrons flashing. We float around the molecules, examining their regularity from all angles and, incredibly, see atoms arranged like Kepler's pomegranate. This is where the snowflakes' symmetry begins. Hexagonal rings of water molecules build on one another, repeating the six-sided rhythm over and over, magnifying the arrangement of oxygen atoms to a scale visible to human eyes.

The basic hexagonal shape of snowflakes is elaborated in varied ways as the ice crystal grows, with the temperature and humidity of the air determining the final shape. Hexagonal prisms form in very cold, dry air. The South Pole is covered with these simple forms. As temperatures rise, the straightforward hexagonal growth of ice crystals starts to destabilize. The cause of this instability is still not fully understood, but it seems that water vapor freezes faster on some ice crystal edges than others, and the speed of this accretion is strongly affected by slight variations in air conditions. In very wet air, arms sprout from the snowflakes' six corners. These arms then turn into new hexagonal plates or, if the air is warm enough, they grow yet more appendages, multiplying the arms of the growing star. Other combinations of temperature and humidity cause the growth of hollow prisms, needles, or furrowed plates. As snowflakes fall, the wind tosses them through the air's innumerable slight variations of temperature and humidity. No two flakes experience exactly the same sequence, and the particularities of these divergent histories are reflected in the uniqueness of the ice crystals that make up each snowflake. Thus, the chance events of history are layered over the rules of crystal growth, producing the tension between order and diversity that so pleases our aesthetic sense.

If Kepler could visit us today he would perhaps be pleased by our solution to the puzzle of the snowflake's beauty. His insights into the arrangement of pomegranate seeds and honeybee cells were on the right track. The geometry of stacked spheres is the ultimate cause of the snowflake's shape. But because Kepler knew nothing of the atomic basis of the material world, he could not imagine the minute oxygen atoms from which ice's geometry grows. However, in a roundabout way, Kepler contributed to the solution to the problem. His musings on the snowflake prompted other mathematicians to investigate the geometry of packed spheres, and these studies contributed to the development of our modern understanding of atoms. Kepler's essay is now regarded as one of the foundations of modern atomism, a worldview that Kepler himself explicitly rejected when he told a colleague that he could not go *ad atomos et vacua*, "to the atoms and the void." Kepler's insights helped others to see what he could not.

I examine again the glassy stars on my fingertip. Thanks to Kepler and those who followed him, I see not just snowflakes but sculptures of atoms. Nowhere else in the mandala is the relationship between the infinitesimally small atomic world and the larger realm of my senses so simple. Other surfaces here—rocks, bark, my skin and clothes—are made from complicated tangles of many molecules, so my view of them tells me nothing straightforward about their minute structure. But the form of the six-sided ice crystals gives a direct view of what should be invisible, the geometry of atoms. I let them fall from my hand, and they return to the oblivion of massed white.

January 21st—The Experiment

A polar wind rips across the mandala, streaming through my scarf, pushing an ache into my jaw. Not counting the windchill, it is twenty degrees below freezing. In these southern forests such cold is unusual. Typical southern winters cycle between thaws and mild freezes, with deep chills arriving for a few days each year. Today's cold will take the mandala's life to its physiological limits.

I want to experience the cold as the forest's animals do, without the protection of clothes. On a whim, I throw my gloves and hat onto the frozen ground. The scarf follows. Quickly, I strip off my insulated overalls, shirt, T-shirt, and trousers.

The first two seconds of the experiment are surprisingly refreshing, a pleasant coolness after the stuffy clothes. Then the wind blasts away the illusion and my head is fogged with pain. The heat streaming out of my body scorches my skin.

A chorus of Carolina chickadees provides the accompaniment to this absurd striptease. The birds dance through the trees like sparks from a fire, careening through twigs. They rest no more than a second on any surface, then shoot away. The contrast on this cold day between the chickadees' liveliness and my physiological incompetence seems to defy nature's rules. Small animals should be less able to cope with the cold than their larger cousins. The volume of all objects, including animal bodies, increases by the cube of the object's length. The amount of heat that an animal can generate is proportional to the volume of its body, so heat generation also increases with the cube of body length. But the surface area, where heat is lost, increases by only the square of length. Small animals cool rapidly because they have proportionally much more body surface than body volume.

The relationship between the size of animals and the rate of heat loss has produced geographic trends in body sizes. When an animal species exists over a large area, the individuals in the north are usually larger than those in the south. This is known as Bergmann's rule, after the nineteenth-century anatomist who first described the relationship. Carolina chickadees in Tennessee live toward the northern end of the species' range, and they are ten to twenty percent larger than individuals from the southern limit of the range in Florida. Tennessee birds have tipped the balance between surface area and body volume to match the colder winters here. Farther north, Carolina chickadees are replaced by a closely related species, the black-capped chickadee, which is ten percent larger again.

Bergmann's rule seems remote as I stand naked in the forest. The wind gusts hard and the burning sensation in my skin surges. Then, a deeper pain starts. Something behind my conscious mind is trapped and alarmed. My body is failing after just a minute in this winter chill. Yet, I weigh ten thousand times more than a chickadee; surely these birds should be extinguished in seconds.

The chickadees' survival depends, in part, on their insulating feathers, which give them an advantage over my naked skin. The smooth upper layer of plumage is plumped by hidden downy feathers. Each down feather is made from thousands of thin protein strands. These tiny hairs combine to make a lightweight fuzz that holds heat ten times better than the same thickness of coffee-cup Styrofoam. In the winter, birds increase by fifty percent the number of feathers on their bodies, adding insulative power to their plumage. On cold days, muscles at the base of the feathers tense, puffing the bird and doubling the thickness of the insulation. Yet all this impressive protection merely slows the inevitable. Chickadee skin does not burn in the cold like mine, but heat still courses out. A centimeter or two of downy fluff buys just a few hours of life in the extreme cold.

I lean into the wind. The sense of alarm builds. My body shakes in uncontrollable spasms.

My usual heat-generating chemical reactions are now totally inadequate, and my muscles' shivering paroxysms are the last defense against a falling core temperature. Muscles fire seemingly randomly, pulling against one another so that my body shudders. Inside, food molecules and oxygen are burned, just as they are when muscles cause me to run or lift, but now this burn produces a rush of heat. The violent shuddering of my legs, chest, and arms warms the blood, which then carries heat to the brain and the heart.

Shivering is also the chickadees' main defense against the cold. Throughout the winter, the birds use their muscles as heat pumps, shivering whenever the temperature is cold and the birds are not active. Slabs of flight muscle in the chickadees' chests are the primary sources of heat. Flight muscles account for about a quarter of a bird's body weight, so shivering produces great surges of hot blood. Humans have no comparably huge muscles in our bodies, so our experiences of shivering are weak in comparison.

As I stand shaking, fear surfaces. I panic and dress as fast as I can. My hands are numb, and I grasp my clothes with difficulty, fumbling with zippers and buttons. My head aches as if my blood pressure has suddenly soared. My only desire is to move quickly. I walk, jump, and wave my arms. My brain signals: make heat, fast.

The experiment has lasted only a minute, just one ten-thousandth of the duration of this week of arctic air. Yet my physiology reels. My head pounds, my lungs can't grasp enough air, and my limbs seem paralyzed. Had the experiment continued minutes longer my core body temperature would have dropped into hypothermia. Muscle coordination would have fled, then sleepiness and hallucinations would have taken over my mind. Human bodies normally keep themselves at about thirty-seven degrees Celsius. If the body temperature drops just a few degrees, to thirty-four, mental confusion sets in. At thirty degrees, organs start to shut down. In cold winds like today's, these few degrees of temperature loss can take place in just an hour of naked exposure. Stripped of my clever cultural adaptations to the cold, I'm revealed as a tropical ape, profoundly out of place in the winter forest. The chickadees' insouciant mastery of this place is humbling.

After I've waved and stamped my limbs for five minutes, I huddle down into my clothes, still shaking but no longer panicked. My muscles feel tired and I'm winded, as if I've just sprinted. I'm feeling the aftereffects of the exertion required for heat generation. When shivering continues for more than a few minutes, it can rapidly deplete an animal's energy reserves. For both human explorers and wild animals, starvation is often the prelude to death. As long as food supplies last, we can shiver and cling to life, but we cannot survive with empty stomachs and drained fat reserves.

I will replenish my reserves when I retreat to my warm kitchen, drawing on the winter-defying technologies of food preservation and transportation. But chickadees have no dried grains, farmed meat, or imported vegetables. Survival in the winter forest demands that chickadees uncover enough food to fuel their four-pennyweight furnaces.

The energy used by chickadees has been measured both in the laboratory and in free-living birds. On a winter day, the birds need up to sixty-five thousand joules of energy to keep themselves alive. Half this energy is used to shiver. These abstract measures become more understandable when they are converted into the currency of bird food. A spider the size of a comma on this page contains just one joule. A spider that fits within a capitalized letter holds one hundred joules. A word-sized beetle has two hundred and fifty joules. An oily sunflower seed has more than one thousand joules, but the mandala's birds have no seed-filled bird feeder. Chickadees must daily find hundreds of food morsels to meet their energy budget. Yet the mandala's larder looks utterly empty. I see no beetles, spiders, or food of any kind in the ice-blasted forest.

Chickadees can coax sustenance out of the seemingly barren forest in part because of their outstanding eyesight. The retinas at the back of the chickadees' eyes are lined with receptors that are two times more densely packed than are mine. The birds therefore have high visual acuity and can see details that my eyes cannot. Where I see a smooth twig, birds see a fractured, flaking contortion, pregnant with the possibility of hidden food. Many insects pass the winter ensconced inside tiny cracks on tree bark, and the chickadees' discerning eyes uncover these insect hideaways. We can never fully experience the richness of this visual world, but peering through a magnifying lens gives us an approximation. Details that are normally invisible snap into view. Chickadees spend most of their winter days passing their superior eyes over the forest's twigs, trunks, and leaf litter, sleuthing concealed food.

Chickadee eyes also perceive more colors than mine can. I view the mandala with eyes that are equipped with three types of color receptor, giving me three primary colors and four main combinations of primary colors. Chickadees have an extra color receptor that detects ultraviolet light. This gives chickadees four primary colors and eleven main combinations, expanding the range of color vision beyond what humans can experience or even imagine. Bird color receptors are also equipped with tinted oil droplets that act as light filters, allowing only a narrow range of colors to stimulate each receptor. This increases the precision of color vision. We lack these filters, so even within the range of light visible to humans, birds are better able to discriminate subtle differences in color. Chickadees live in a hyperreality of color that is inaccessible to our dull eyes. Here in the mandala, they use these abilities to find food. Ultraviolet light reflects from the dried wild grapes that are sparsely scattered across the forest floor. Wings of beetles and moths are sometimes tinged with ultraviolet, as are some caterpillars. Even without the advantage of ultraviolet vision, insect camouflage is unmasked by slight imperfections detected by the birds' precise perception of color.

The visual abilities of birds and mammals differ because of events in the Jurassic, one hundred and fifty million years ago. At that time, the lineage that gave rise to modern birds split from the rest of the reptiles. These ancient birds inherited the four color receptors of their reptilian ancestors. Mammals also evolved from reptiles, splitting away earlier than the birds. But, unlike birds, our protomammal ancestors spent the Jurassic as nocturnal shrewlike creatures. Natural selection's shortsighted utilitarianism had no use for sumptuous color in these night-dwelling animals. Two of the four color receptors that the mammals' ancestors bequeathed to them were lost. To this day, most mammals have just two color receptors. Some primates, including those that gave rise to humans, later evolved a third.

The chickadees' acrobatic bodies let them put their vision to good use. A wing-flick takes a bird from one branch to another. Feet grasp a twig, then the bird falls, swinging from a branch tip. The beaks probe as the bird's body pivots, still hanging, then wings flash open and the bird flits up to another small twig. No surface is left unexamined. The birds spend as much time upside down, peering under twigs, as they do upright.

Despite the vigor of their search, the chickadees catch no prey while I watch. Chickadees, like most birds, give a distinctive backward flick of their heads as they swallow or, if they find a bigger morsel, will hold the food in their feet as they pound it with their beaks. The flock stays in my sight for just fifteen minutes, finding no food. The chickadees may need to call on their fat reserves to survive the cold. These reserves are essential to winter survival, and they allow chickadees to make good use of winter's variability. When the weather warms, or when birds find a cluster of spiders or berries, the flush of food is turned into fat that carries the birds through times when the feeding is poor and the weather is cold.

The degree of fatness varies among individual birds. Chickadees feed in socially stratified flocks, usually composed of a dominant pair and several subordinates. Dominant birds get access to whatever food the flock finds, so they generally eat well whatever the weather. These high-ranking birds have trim bodies. Subordinate chickadees bear the brunt of winter's hardship, eating well only intermittently. The low-status birds, often youngsters or failed breeders, compensate for the variability of their food intake by getting fatter, buying insurance against lean times. But there is a cost to chickadee fatness. Rotund birds are easier prey for hawks. The fatness of each chickadee is a balance between the risk of starvation and the risk of predation.

Chickadees supplement their fat stores by jabbing insects and seeds under flaking bark, storing food for later recovery. Carolina chickadees are particularly fond of caching food by poking it into the undersides of small branches. This habit may be a guard against thievery from less agile bird species. Nonetheless, caches are vulnerable to plunder, so each chickadee flock in the forest defends a winter territory from which neighbors are vigorously excluded. Non-caching chickadees in other parts of the world are much less territorial.

Larger bird species often join chickadee flocks in the winter. Today, a downy woodpecker chisels for larvae in the bark of an oak tree, then flies after the chickadees when they flit east. A tufted titmouse also travels with the flock. The titmouse bounces among branches like the chickadees, but it is less agile, preferring to light on twigs without swinging from branch ends. All the birds call, keeping the flock together. The chickadees and titmouse chatter and whistle, the woodpecker gives high-pitched *pik* notes. This flocking behavior gives the group members safety from hawks, which are easier to spot when many eyes are vigilant. But chickadees pay a price for safety in the crowd. Tufted titmice are twice as heavy as chickadees, and the larger birds dominate, pushing the chickadees away from dead branches, higher twigs, and other preferred feeding locations. These subtle changes in location result in significant lost feeding opportunities for chickadees. In flocks where titmice are absent, chickadees are better fed. Survival in the winter mandala therefore requires not just sophisticated physiology but careful negotiation of social dynamics.

Daylight is fading now. I move my chilled limbs and rub my ice-crusted eyes in preparation for the walk out of the forest. The birds will continue their search for food a few minutes longer, then they will head to their roosts. As light fails and the temperature drops, chickadees will gather in holes left by fallen tree limbs, sheltering from the wind's heat-ripping power. The birds huddle in groups, giving a nod to Bergmann's rule by creating a ball of birds with a large volume and a relatively small surface area. Then the chickadees' body temperature will fall by ten degrees into an energy-saving hypothermic torpor. At night, as in the day, integrated behavioral and physiological adaptations give the birds an edge over winter. Torpor combined with huddling halves the chickadees' nighttime energy needs.

The chickadees' adaptations to the cold are remarkable, but they are not always adequate. There will be fewer chickadees in the forest tomorrow. Winter's chill hands will pull down many of these birds, dragging them deeper than the appalling emptiness I felt when I experienced the cold. Only half the chickadees that fed among the falling autumn leaves will live to see the oak buds open in the spring. Nights such as tonight cause most of the birds' winter mortality.

This week's glacial temperatures will last just a few days, but the spike in bird mortality will change the forest in ways that extend throughout the year. Deaths on winter nights check the chickadee population, trimming any birds that exceed the scant supply of winter food. Carolina chickadees each require, on average, three or more hectares of forest to sustain themselves. This square meter of mandala therefore supports just a few hundred-thousandths of a chickadee. Tonight's cold will remove any excess.

When summer arrives, the mandala will be able to support many more birds. But because the abundance of resident species like chickadees is kept low by winter's meager supplies, the food available in summer vastly exceeds the resident birds' appetites. This great seasonal flush of food creates an opportunity that is exploited by migrant birds that risk long flights from Central and South America to feed on the excess in forests throughout North America. Winter's cold is therefore responsible for the annual migration of millions of tanagers, warblers, and vireos.

Overnight deaths will also fine-tune the chickadee species' fit to its environment. Smaller Carolina chickadees will be more likely to perish than their bulkier kin, reinforcing Bergmann's latitudinal pattern. Likewise, extreme cold will purge from the population those birds whose shivering abilities, feather fluffiness, or energy stores are deficient. In the morning, the chickadee population in this forest will be better matched to winter's demands. This is natural selection's paradox: from death comes life's increasing perfection.

My own physiological inadequacy in the cold also has its origins in natural selection. I am out of place at the icy mandala because my ancestors have dodged selection for cold-hardiness. Humans evolved from apes that lived for tens of millions of years in tropical Africa. Keeping cool was a much bigger challenge than keeping warm, so we have few bodily defenses against extreme cold. When my ancestors left Africa for northern Europe, they brought with them fire and clothes, carrying the tropics to the temperate and polar regions. This cleverness produced less suffering and fewer deaths, unquestionably good outcomes. But comfort sidestepped natural selection. We are condemned by our skill with fire and cloth to be forever out of place in the winter world.

Darkness comes and I retreat toward my inheritance, the warm hearth, leaving the mandala to the avian masters of the cold. This mastery was earned the hard way, through thousands of generations of struggle. I wanted to experience the cold as the mandala's animals do, but I now realize that this was impossible. My experiences come through a body that has taken a different evolutionary path from that of the chickadees, precluding any fully shared experience. Despite this, my nakedness in the cold wind has deepened my admiration for these others. Astonishment is the only proper response.

January 30th—Winter Plants

A continuous low-pitched roar comes from the wind raking the trees on the high bluff above the mandala. Unlike the northerly gales earlier in the week, this wind is from the south, and the bluff protects the mandala from all but a few eddies and gusts. The changing wind has eased the temperature. It is just a couple of degrees below freezing, warm enough to sit comfortably for an hour or more in winter clothes. The urgent, unrelenting physical pain of the cold is gone, and my body welcomes the benign air with a glow of quiet pleasure.

Birds in a passing flock seem to revel in their release from the arctic death grip. Five species travel together: five tufted titmice, a couple of Carolina chickadees, one Carolina wren, a golden-crowned kinglet, and a red-bellied woodpecker. The flock seems bound by invisible elastic threads; when a single bird gets left behind or strays beyond the ten-meter radius of the flock, it is yanked back to the center. The whole flock is a rolling ball of agitation as it moves through the inanimate snowy forest.

The titmice are the most vocal of the birds, streaming a continual jumble of sound. Each titmouse jabs high-pitched *seet* notes, creating an irregular beat against which play their other calls, hoarse whistles and squeaks. Some birds repeat *pee-ta pee-ta,* a sound that was absent from their repertoire earlier this week in the deep cold. This bright two-noted sound is the breeding song. Despite the snow, these birds are already turning their attention to spring. Egg laying will not happen for another couple of months, but the extended social negotiations of courtship have begun.

The life-filled exuberance of the birds contrasts with the mandala's plants. Their gray branches and bare twigs below offer a scene of desolation. Death juts from the snow: fallen, partly decayed maple branches and the frayed stubs of leafcup stems protrude, each stem ringed by a circle of sublimated snow that reveals the dark litter below. Winter appears to have delivered a thorough defeat.

Yet life persists.

Bare shrubs and trees are not the skeletons they appear to be. Every twig and trunk is wrapped by living tissue. Unlike birds that survive by fighting the cold with food won from winter's tight fist, plants somehow endure without re-creating an internal summertime. The birds' survival is astonishing, but the plants' resurrection after a full surrender is so far removed from human experience that it edges into scandal. The dead, especially the frozen dead, should not return.

But, return they do. Plants survive in the same way that a sword swallower survives—with careful preparation and meticulous attention to sharp edges. Plant physiology can generally cope with mere chilling. Unlike the chemical reactions that sustain humans, plant biochemistry can run at many different temperatures, and it does not fail when cooled. But when cooling turns to freezing, problems start. Expanding ice crystals will puncture, tear, and destroy the delicate inner architecture of cells. Plants in winter must swallow tens of thousands of blades, keeping each one away from their fragile hearts.

Plants start their preparations several weeks ahead of the first freezes. They move DNA and other delicate structures to the centers of their cells, then wrap them in cushioning. The cells get fattier, and the chemical bonds in these fats change shape to make them fluid in cold temperatures. The membranes around cells become leaky and flexible. The transformed cells are padded and limber, able to absorb ice's violence without harm.

Preparations for winter take days or weeks to complete. A frost out of season will kill branches that, when properly acclimated, could stand the coldest nights of the year. Native plant species are seldom caught out by freezes; natural selection has taught them the seasonal rhythms of their homes. But exotic plants have no local knowledge and are often heavily pruned by winter.

Cells not only change their physical structure but soak themselves in sugars, lowering the freezing point like salt scattered on icy roads. Sugaring happens only inside the cells, leaving water around the cells unsweetened. This asymmetry allows plants to exploit an expected gift from the laws of physics: heat is released when ice forms. Cells surrounded by freezing water receive a temperature boost of several degrees. During the first frosts of winter, the sugared insides of cells are protected by the unsugared water surrounding them. Farmers exploit this burst of warmth by misting their crops on frosty nights, adding another layer of heat-releasing water.

Once all the water between cells has frozen solid, no more heat is released. But water inside cells is still liquid. This liquid oozes out of the leaky membrane around the cell. As the water leaks out, it leaves behind the sugars, which, being large molecules, cannot pass through the membrane. This process gradually draws water out of the cell as temperatures drop, increasing the inner concentration of sugars and further dropping the freezing point. When temperatures are very low, cells pucker into balls of syrup, unfrozen repositories of life, surrounded by shards of ice.

The Christmas fern and the mosses in the mandala face an additional challenge. Although their evergreen leaves and stems let them feed on warm winter days, the source of their green, chlorophyll, can be unruly in cold weather. Chlorophyll captures energy from the sun and converts it into a buzz of excited electrons. In warm weather, the electrons' energy is quickly shunted to the food-making process in the cell. But this shunt seizes up in cold weather, leaving cells awash in overexcited electrons. Unchecked, the undirected energy will trash the cell. To forestall the riot of electrons, evergreens prepare for winter by stocking their cells with chemicals that intercept and neutralize the unwanted electron energy. We know these chemicals as vitamins, particularly vitamins C and E. Native Americans also knew this and chewed winter evergreens to keep healthy through the winter.

Ice permeates the mandala's plants, but each cell carefully recoils, enforcing a microscopic separation between ice and life. By reversing this cellular contraction, twigs, buds, and roots are able to revive in spring and carry on almost as if winter had not happened. A few plant species, however, take a different path. Leafcup herbs completed their short eighteen-month lives last fall and now stand dead, surrendered entirely to winter. They have sublimated into a new physical form, like snow passing into vapor. Like vapor, these new forms are invisible, but they surround me. Buried in the mandala's litter are thousands of leafcup seeds, waiting out winter. Because seeds have hard coats and dry interiors, they pass through the cold months largely protected from the assaults of ice.

The impression of desolation in the mandala is superficial. Within the bounds of this one square meter are hundreds of thousands of plant cells, each one wrapped into itself, intensified in its withdrawal. The quiet gray exterior of plants, like gunpowder, belies the energy that is latent here. So, although titmice and other birds give a vigorous display of life in January, they are trifles compared to the power stored in the quiescent plants. When spring sparks the mandala, the energy released will carry the whole forest, birds included, through another year.

February 2nd—Footprints

The tips of a maple-leaf viburnum have been chiseled off, leaving beveled stubs along the shrub's branches. The animal that clipped these tender shoots has left three footprints in the mandala, aligned east to west. Two almond-shaped impressions make up each footprint, sunk two inches into the leaf litter. This is the signature of a cloven hoof, the seal of the artiodactyl clan. Like nearly every terrestrial community the world over, the mandala has been browsed by a cleft-hoofed mammal, in this case a white-tailed deer.

The deer that passed through the mandala last night was careful in its choice of browse. The viburnum shrub had stored food in branch tips, readying itself for spring. These young tips were not yet toughened and woody. The shrub's tender growth has now been robbed, digested, and reinvested in deer muscle or, if the nibbler was a doe, in the body of a fawn in her womb.

The deer had help. Freeing the food locked inside the tough cells of twigs and leaves requires a partnership between the very large and the very small. Big multicellular animals can nip off and chew woody material, but they cannot digest cellulose, the molecule that constitutes most plant matter. Microbes, tiny single-celled organisms such as bacteria and protists, are physically puny but chemically powerful. Cellulose does not give them pause. Thus is born a gang of thieves: animals that walk around and grind up plants, paired with microbes that digest pulverized cellulose. Several groups of animals have independently developed this plan. Termites work with protists in their gut; rabbits and their kin harbor microbes in a large chamber at the end of their gut; the hoatzin, an improbable leaf-eating bird from South America, has a fermentation sac in its neck; ruminants, including deer, have a huge bag of helpers in a special stomach, the rumen.

Microbial partnerships allow large animals to use the vast stores of energy locked up in plant tissues. Those animals, including humans, that have not entered into a deal with microbes are limited to eating soft fruits, a few easily digestible seeds, and the milk and flesh of our more versatile animal cousins.

The saplings in the mandala were pinched between the deer's lower teeth and the tough pad on its upper jaw that takes the place of upper front teeth. The woody morsels were sent to the back teeth to be ground up, then swallowed. When these pieces hit the rumen they entered another ecosystem, a huge churning vat of microbes. The rumen is a sac that branches off the rest of the deer's gut. All food, except the mother's milk, is sent to the rumen before it can move into the rest of the stomach, then on to the intestines. The rumen is surrounded by muscles that churn the contents. Flaps of skin inside the rumen act like baffles in a washing machine, flipping the food over as it moves.

Most microbes in the rumen cannot live in the presence of oxygen. They are descendants of ancient creatures that evolved in a very different atmosphere. Only when photosynthesis was invented, about two and a half billion years ago, did oxygen become part of earth's air and, because oxygen is a dangerous, reactive chemical, this poisoning of the planet wiped out many creatures and forced others into hiding. These oxygen-haters live to this day in lake bottoms, in swamps, and deep in the soil, eking out an existence in oxygen-free environments. Other creatures adapted to the new pollutant and, using an elegant sidestepping maneuver, turned the toxic oxygen to their advantage. Thus was born respiration using oxygen, an energy-liberating biochemical trick that we have inherited. Our lives therefore depend on an ancient form of pollution.

The evolution of animal guts gave the oxygen-hating refugees a potential new place in which to hide. Not only are guts relatively free of oxygen, they also have every microbe's dream: a continual supply of minced food. But there was a problem. Animal stomachs are generally full of acidic digestive juices designed to tear apart living tissue. This prevented most animals from harboring plant-digesting microbes. However, the ruminants changed their stomachs, mastering the hotelier's art, and they have been rewarded by a four-star rating of evolutionary success. The centerpiece of this hospitality is the position and friendliness of the rumen, which comes before the rest of the gut and is kept neutral, neither acid nor alkaline. Microbes thrive in this churning spa. The animal's saliva is alkaline, so the acidic products of digestion are neutralized. Any incoming oxygen is soaked up by a small team of bacterial chambermaids.

The rumen functions so well that scientists equipped with the most sophisticated test tubes and vats have not been able to replicate, let alone beat, the growth rate or digestive prowess of the rumen's microbes. The rumen's performance is due to the exquisite biological complexity that thrives in its pampered chambers. A million million individual bacteria of at least two hundred species swim through every milliliter of rumen fluid. Some of these microbes have been described; others await description or discovery. Many of the microbes are found only in rumens, presumably having diverged from their free-living ancestors during the fifty-five million years that have passed since the rumen's origin.

Within the rumen, the bacterial proletariat is preyed upon by a bevy of protists, all of which are single-celled but hundreds or thousands of times bigger than the bacteria. Fungi parasitize these protists, infecting then bursting the fat cells. Other fungi float free in the rumen fluid or colonize scraps of plant material. The diversity of life in the rumen makes possible the complete digestion of the plant remains. No single species can fully digest a plant cell. Each species takes a small part of the overall process, chopping up its favorite molecules, harvesting the energy it needs to grow, and then sending back its wastes to the rumen fluid. These wastes become another creature's food, building a cascading web of disassembly. Bacteria destroy most of the cellulose, aided by some fungi. Protists have a special fondness for starch grains, perhaps regarding them as potatoes to accompany their meal of bacterial sausages. Nutrients in the rumen are passed up a miniature food web, then released back into the rumen's fluid, mimicking the nutrient cycles of larger ecosystems. The deer's belly contains a mandala of its own, an intricate dance of lives, sustained by hungry lips and teeth. Young ruminants must build their rumen community from scratch, a process that takes several weeks. During this time, they nibble their mother, the soil, and the vegetation, gathering and swallowing the microbes that will become their helpers.

The rumen's ecosystem is a self-sacrificing mandala, embodying endless change. Microbes are carried out of the rumen along with digested plant cells. They travel to the second part of the deer's stomach and are swamped with acid and digestive juices. For these microbes, the gut's hospitality has ended. The innkeeper kills and digests them, pocketing their proteins and vitamins along with the liquefied plant remains.

The rumen retains plant solids and the microbes that cling to them, ensuring both the complete digestion of the plant and the continuity of the rumen's microbial community. The deer hastens the breakdown of these solids by bringing them back into its mouth, chewing the cud, then swallowing the pulverized remains. This rumination allows the deer to "wolf" down its food, literally on the hoof, then chew it in a safe hiding place, away from real wolves.

As the seasons change, the deer's browsing moves from one part of the plant to another. The woody food of winter will change to springtime greenery, then autumn acorns. The rumen adapts to these changes through the gradual waxing and waning of the members of its community. Bacteria suited to digestion of soft leaves increase through the spring, then taper away in winter. No top-down control by the deer is needed to direct this change; competition among the rumen inhabitants automatically matches the rumen's digestive capabilities to the food available. But sudden changes in diet can disrupt this elegant molding of the rumen community to its environment. If a deer is fed corn or leafy greens in the middle of winter, its rumen will be knocked off balance, acidity will rise uncontrollably, and gases will bloat the rumen. Indigestion of this kind can be lethal. Young ruminants face a similar digestive problem when they suckle their mother's teats. Milk would ferment and create gas in the rumen, especially in immature animals whose rumens have yet to be fully colonized by microbes. The sucking reflex therefore triggers the opening of a bypass that sends milk past the rumen, into the next part of the stomach.

Nature seldom throws rapid dietary change at ruminants, but when humans feed domesticated cows, goats, or sheep, they must address the rumen's needs. These needs do not necessarily conform to the desires of human commodity markets, so the rumen's balance is the bane of industrial agriculture. When cows are taken from pasture and suddenly confined to feedlots to be fattened on corn, they must be medicated to pacify the rumen community. Only by stamping down the microbial helpers can we try to impose our will on the cow's flesh.

Fifty-five million years of rumen design versus fifty years of industrial agriculture: we face questionable odds.

The deer's effects in the mandala were subtle. At first glance, shrubs and saplings appear unmolested. Only close observation reveals the missing tips of branches and the short, amputated stubs of side shoots. About half the dozen shrub stems in the mandala have been trimmed, but none of them have been cut back to stumps. I infer that deer and their microbial companions are frequent visitors to the mandala, but the deer are not starving. They can afford to nibble the succulent ends of twigs, leaving the woody stems behind. This choosiness is becoming a threatened luxury among white-tailed deer in the eastern forests. Across much of the deer's range plant defenses are deployed in vain: deer populations have expanded rapidly, and the teeth and rumens of these growing hordes have sterilized the forest of saplings, shrubs, and wildflowers.

Many ecologists claim that the recent growth of the deer population is a continentwide catastrophe. Equivalent, perhaps, to throwing corn into a winter rumen; the community is thrown into an unnatural disequilibrium. The case against the deer seems unassailable. Deer numbers are growing. Plant populations are in decline. Shrub-nesting birds cannot find nest sites. Tick-borne diseases lurk on suburban lawns. We have eliminated predators, first Native Americans, then wolves, then modern hunters, whose numbers dwindle each year. Our fields and towns have cut the forest into ribbons and rags, creating the edge habitat in which deer love to feed. We have carefully nurtured deer populations with game-protection laws that time the hunting season to have the smallest possible effect on the deer population. Surely the forest's viability is endangered?

Perhaps, but a longer view adds some mists of uncertainty to this black-and-white portrait of the role of deer in the eastern forests. Our cultural and scientific memories of what a "normal" forest should look like arose at a peculiar moment in history, a moment when deer, for the first time in millennia, had been extirpated from the forest. Large-scale commercial hunting in the late nineteenth century edged the deer population toward extinction. Deer were eliminated from most of Tennessee, including from this mandala. No deer visited the mandala between 1900 and the 1950s. Then, releases of deer transplanted from elsewhere, combined with elimination of bobcats and feral dogs, gradually pushed the population of deer upward until the 1980s, when deer were once again abundant. A similar pattern was replicated across the eastern forest.

This history distorts our scientific understanding of the forest.

Most of the scientific studies of eastern North American forest ecology in the twentieth century were conducted in an abnormally unbrowsed forest. This is especially true of the older studies that we use as a benchmark to measure ecological change. The benchmark is misleading: at no other time in the history of these forests have ruminants and other large herbivores been absent. Our memory, therefore, recalls an abnormal forest, limping along without its large herbivores.

Disquieting possibilities grow out of this history. Wildflowers and shrub-nesting warblers may be experiencing the end of an unusual era of ease. "Overbrowsing" by deer may be returning the forest to its more usual sparse, open condition. The surviving diaries and letters of early European settlers lend some support to these ideas. Thomas Harriot wrote from Virginia in 1580 that "of . . . deare, in some places there are great store"; Thomas Ashe reports in 1682 that "there is such infinite herds, that the whole country seems but one continued park"; Baron de La Hanton continues this theme in 1687: "I cannot express what quantities of deer and turkeys are to be found in these woods."

The writings of these European colonists are suggestive but hardly definitive. Their letters may be biased by boosterism for the colonists' project, and they were entering a continent whose human occupants, most of whom were hunters, had just been decimated by disease and genocide. But the stories of genocide survivors and the archaeological evidence left by their ancestors suggests that deer were abundant even before the Europeans arrived. Native Americans cleared and burned forests to encourage the growth of young vegetation that, in turn, fired the deer's fecundity. Deer meat and hides made human life possible in the winter, and deer spirits danced through the mythology of these first human inhabitants of the Americas. Historical and archaeological information therefore all points to the same conclusion: deer were plentiful inhabitants of our forests before guns removed them in the 1800s. The deerless forests of the early and middle 1900s were aberrations.

The case against our modern deer phobia deepens when we look back beyond the arrival of humans on this continent. Temperate forest has been growing in eastern North America for the last fifty million years. In those ancient times, the forest grew in a thick band across Asia, North America, and Europe. This swath was cut into fragments by the cooling of the earth's climate, especially by the periodic ice ages that pushed temperate forests southward, then drew them northward as the ice retreated. Now, the remnants of this forest grow in widely separated patches in eastern China, Japan, Europe, the Mexican highlands, and eastern North America. The temperate forest's dance across the continents has one unvarying theme: the presence of mammalian browsers, often in large numbers.

The deer that walked across the mandala is one of the last representatives of a much larger plant-trimming bestiary. Giant ground sloths hulked their rhino-sized bodies around the forest, browsing vegetation. They were accompanied by woodland musk oxen, giant herbivorous bears, long-nosed tapirs, peccaries, woodland bison, several species of extinct deer and antelope, and, most dramatic of all, the mastodons. The mastodons were relatives of the modern elephant, with tusks and a broad, low head. They stood three meters tall at the shoulder and browsed through the northern edge of the eastern forest. They, like many other large herbivores, went extinct at the end of the last ice age, about eleven thousand years ago. The ice ages had come and gone before, but this thaw brought with it a new predator, humans. Shortly after the arrival of humans, most of the large herbivores were gone. The smaller mammals were minimally affected by this extirpation; only large, meaty creatures disappeared.

Fossil evidence of these large herbivores abounds in caves and swamps across the eastern United States. These fossils provided fuel for the nineteenth-century debate about evolution. Darwin thought these animals were further evidence for the idea that the natural world is always in flux. He commented, "It is impossible to reflect on the state of the American continent without astonishment. Formerly it must have swarmed with great monsters; now we find mere pygmies compared to the antecedent, allied races." Thomas Jefferson disagreed, believing that giant sloths and other creatures must still be alive. After all, why would God create them, then kill them off? Creation reflected God's perfect handiwork, therefore nature would unravel if pieces were allowed to fall away. Jefferson instructed the explorers Lewis and Clark to bring back reports of these creatures from their trek to the Pacific coast. The expedition found no evidence of living mastodons, sloths, or any other extinct creatures. Darwin was right; pieces of creation can be destroyed.

Like the footprints left by the deer's visit to the mandala, the passing herbivores have left signs in the architecture of some of our native plants. Honey locust and holly trees have thorny stems and leaves. These thorns are deployed only to three meters' height, twice as high as any living herbivore can reach but exactly the right height to deter the extinct mega-browsers. The honey locust is doubly lost because its seedpods, which are two feet long, are too large for any living native species to consume whole and thus disperse the seeds, although they are perfectly sized for large extinct herbivores such as mastodons and ground sloths. Osage orange's milky softball is another fruit whose seed-dispersing partner has died. Similar fruits on other continents are eaten by elephants, tapirs, and other large herbivores of the kind that exist only as fossils in North America. These widowed plants wear history on their sleeves, giving us a glimpse into the bereavement of the whole forest.

The structure of ancient forests is forever hidden from us, but the bones of extinct browsers and the stories of the first Americans suggest that this was not an easy place for shrubs and saplings to thrive. North American forests have experienced fifty million years of browse, followed by ten thousand years of drastically reduced mammalian herbivory, then one hundred strange years of no browse at all. Might the ancient forests have been patchy and sparse, kept trimmed by herds of wandering herbivores? Certainly these herbivores had enemies of their own, which are now gone, or almost so. The sabertooth cat and the dire wolf are extinct; the gray wolf, mountain lion, and bobcat are rare. In the western United States, the giant American lion and the cheetah both preyed on the plant browsers. The existence of these many species of large carnivore is further evidence of the abundance of herbivores. Giant cats and wolves need giant herds of food. The only places in the world that sustain large populations of carnivores are well endowed with browsers. After all, carnivore flesh is just plant matter passed up the food web. So, abundant fossil evidence of large predators is strong evidence for heavy browse on plants.

Humans have eliminated some predators but have lately added three new deer-slaying creatures: domestic dogs, immigrant coyotes invading from the west, and automobile fenders. The first two are effective predators of fawns; the latter is the main suburban killer of adults. We face an impossible equation. On one hand, we have the loss of tens of species of herbivores; on the other we have the replacement of one type of predator by another. What level of browse is normal, acceptable, or natural in our forests? These are challenging questions, but it is certain that the lush forest vegetation that grew in the twentieth century was unusually underbrowsed.

A forest without large herbivores is an orchestra without violins. We have grown accustomed to incomplete symphonies, and we balk when the violins' incessant tones return and push against the more familiar instruments. This backlash against the herbivores' return has no good historical foundation. We may need to take the longer view, listen to the whole symphony, and celebrate the partnership between animal and microbe that has been tearing at saplings for millions of years. Good-bye shrubbery; hello ticks. Welcome back to the Pleistocene.

February 16th—Moss

The mandala's surface is a tumult of water, crackling as the clouds fire volleys, pause, then loose more artillery. Battalions of rain blown in from the Gulf of Mexico have assaulted the forest all week. The world seems made of flowing, exploding water.

Mosses exult in the wetness. They arch into the rain, swollen green. Their transformation is remarkable. Last week they hung parched and bleached on the mandala's rock faces, beaten down by winter. No longer. Their bodies have tapped the clouds' energy.

My own wintertime desiccation has created a thirst for wet, green renewal that moves me to a closer look. I lie at the mandala's edge and lean my face to the mosses. They smell of earth and life, and their beauty rises exponentially with nearness. I am greedy for more and pull out a hand lens, pressing my eye against it as I creep closer.

Two types of moss intermingle on the rock face. Without removing them to the laboratory to examine the shape of their cells under a microscope, I cannot definitively identify them, and so I observe them without naming. One species lies in fat ropes, each rope wrapped in closely spaced leaflets. From a distance the stems look like living dreadlocks; a closer view shows the leaflets are arranged in repeating graceful spirals, like green petals repeated over and over. The other species stands erect, its stems branching like miniature spruce trees. The growing tips of both species are green as baby lettuce. Color darkens behind the tips, shading into the olive green of mature oak leaves.

Luminosity dominates this world; each leaf is one cell layer thick, so light dances and flows through the moss, giving it an internal glow. Water, light, and life have united their powers and broken winter's lock.

Despite their verdant vigor, mosses get little respect. Textbooks write them off as primitive holdouts from an earlier time, prototypes that have been superseded by more advanced plants such as ferns and flowering plants. This notion of mosses as evolutionary leftovers fails on several counts. If mosses were backward hicks dying out in the face of superior modernity, we would expect to see fossil evidence of an early period of glory, followed by a slow descent into obscurity. But the scant fossil evidence shows the reverse. Further, fossils of the first primitive land plants bear scant resemblance to the carefully arranged leaflets and elaborate fruiting stalks of modern mosses.

Genetic comparisons corroborate the fossils' story, showing that the plants' family tree split into four main branches, each of which has been separated from the others for nearly five hundred million years. The order in which these branches split is still controversial, but the liverworts, creeping alligator-skinned lovers of stream margins and wet rock faces, may have been the first to diverge. The ancestors of the mosses broke away next, followed by the hornworts that are the closest relatives to ferns, flowers, and their kin. Mosses have evolved their own way of being, a way that is not now, nor ever was, just a halfway house to a "higher" form.

I gaze through my hand lens and see water caught everywhere in the moss. In the angles between leaves and stems, water is caught in curved silver pools, trapped by surface tension. Droplets don't flow, they clasp and climb. Moss seems to have erased gravity and conjured rising snakes of liquid. This is the world of the meniscus, the lip of water that pulls itself up the wall of a glass cup. And moss is all glass edge, an architecture that invites then traps water in its labyrinthine core.

The relationship of moss to water is hard for us to grasp. Our plumbing is internal, all buried pipes and pumps. Trees likewise keep their conduits below the skin. Even our houses are plumbed from within. Mammals, trees, houses: these belong in the world of the very large. The microworld of moss operates under different rules. The electrical attraction between water and plant cell surfaces is a powerful force over short distances, and moss bodies are sculpted to master this attraction, moving and storing water on their complex facades.

Grooves on the surface of stems wick water from the mosses' wet interiors to their dry tips, like tissue paper dipped in a spill. The miniature stems are felted with water-hugging curls, and their leaves are studded with bumps that create a large surface area for clinging water. The leaves clasp the stem at just the right angle to hold a crescent of water. These trapped drops are interconnected by water trapped in woolly hairs and surface wrinkles. Moss bodies are swampy river deltas miniaturized and turned vertical. Water creeps from slough to lagoon to rivulet, wrapping its home in moisture. And when the rains stop, the moss has captured five to ten times as much water on its body as it contains within its cells. Moss carries a botanical camel's hump as it trudges through long stretches of aridity.

Mosses work out of a different architectural textbook from that of trees, but the end result is arguably as complex and certainly as successful at long-term evolutionary survival. But the sophistication of moss design does not end with the transport and storage of water. When the rains began a week ago, they triggered a cascade of physiological changes that made possible today's lush growth. Water first wrapped the desiccated moss, then seeped into the thin wooden walls of each cell and slicked the surface of the dry raisins within. These shriveled balls were dormant living cells, and each one's skin was primed to sponge the rain's gift. The cells swelled, the skin pushed against the wooden wall, and life returned.

The push of thousands of cells plumped the plant and raised the moss out of its winter slackness. At the corners of each leaf, large curved cells ballooned with water and levered the leaves away from the stem's axis, opening spaces to hold water and angling the leaves' faces skyward. The inner concave leaf surfaces hold water. The outer convex surfaces harvest sun and air to make the mosses' food. The rain-induced swelling turned each leaf into both water harvester and sun catcher, root and branch.

Inside the cells, havoc reigned. Inrushing water jumbled the cells' innards. Wetted membranes loosened so fast that some of the cells' contents leaked away. These sugars and minerals were forever lost to the plant, the cost of flexibility. But disorder did not last. Before drying, the moss prudently stocked its cells with repair chemicals. Now that the cells have swollen, these chemicals restore and stabilize the cells' flooded machinery. As soon as the moistened cell regains its balance, it will replenish the supply of repair chemicals. The cell will also infuse itself with sugars and proteins that help pack away machinery when conditions dry out.

Mosses are thus equipped at all times to cope with either drought or flood. Most other plants take a more relaxed approach to emergency preparedness, building their rescue kit from scratch when times get hard. This kit building takes time, so rapid drying or wetting will kill the laggards but not the mosses.

Careful preparations are not the only way that mosses overcome drought. They can endure extremes of aridity that would crisp and destroy the cells of other plants. By loading their cells with sugar, dry mosses crystallize into rock candy, vitrifying and preserving the cells' innards. Desiccated moss would be tasty were it not for the fibrous coating and bitter seasoning of the candied cells.

Half a billion years of life on land have turned mosses into expert choreographers of water and chemistry. The lush thickets of moss over the mandala's rocks illustrate the advantages of a limber body and nimble physiology. The surrounding trees, shrubs, and herbs still wear winter's chains, but mosses are unshackled and free to grow. Trees cannot make use of the early thaw. Later, the tables will turn, and trees will use their roots and internal plumbing to dominate the mandala's summer, shading the rootless mosses below. But for now, the trees are paralyzed by their hulking size.

The mosses' late winter eagerness produces benefits that extend beyond their own growth. Life downstream from the mandala profits from the mosses' hold on water. The rainstorm's kinetic energy rakes the hillside, yet the water streaming off the mandala is clear. There is no hint of the mud and silt that bleeds from the fields and towns around. Mosses and the thick forest leaf litter sponge moisture and slow scouring raindrops, turning the artillery assault on the soil into a caress. As the water flows down the mountain, the soil is held in place by a weave of herb, shrub, and tree roots. Hundreds of species work the loom, interpenetrating their warp and weft, turning out a tough, fiber-filled denim that rain cannot tear. By contrast, fields of young wheat and suburban lawns have sparse, loose-woven roots that cannot hold the soil.

The mosses' contributions go beyond acting as the first line of defense against the eroding power of water. Because they have no roots, mosses harvest water and nutrients from the air. Their rough surfaces trap dust and can snatch a healthy dose of minerals from a breath of wind. When the wind carries acidity from tailpipes or toxic metals from power plants, mosses welcome the junk with wet, open arms and draw the pollution into themselves. The mandala's mosses thus cleanse the rain of industrial detritus, clasping and holding heavy metals from car exhaust and the smoke of coal-fired power stations.

When the rain departs, the mosses' sponginess retains water, then slowly releases it. Forests therefore nurture life downstream from themselves, shielding rivers from sudden muddy surges and sustaining flow during dry spells. Evaporation from the wet forest creates clouds of humidity and, if the forest is large enough, generates its own rain. We usually take these gifts without consciousness of our dependence, but economic necessity sometimes jolts us out of our sleep. New York

City decided to protect the Catskill Mountains rather than pay for a man-made water purification plant. The millions of mossy mandalas in the Catskills were cheaper than a technological "solution." In some watersheds in Costa Rica, downstream water users pay upstream forest owners for the service provided by the forested land. Thus the human economy becomes modeled on the reality of the natural economy, and the incentive to tear up the forest is reduced.

In the mandala, the rain continues its pounding. From where I sit, I hear two streams roaring, one on either side of the mandala, both at least a hundred meters away. The rain's volume has turned the sound of their usual quiet trickle to a thundering roil. After an hour or more of huddling in my waterproof clothes, I feel oppressed by the incessant violence. But the mosses seem more at home than ever. Five hundred million years of evolution has given them mastery of wet days.

February 28th—Salamander

A leg flashes across a crevice in the leaf litter. The stub of a tail follows and then disappears into layers of wet leaves. I resist the urge to peel away the leaves; instead I wait, hoping that the salamander will surface again. Several minutes later, a shining head thrusts out, and the salamander sprints into the open. It pushes down another hole, reappears, bursts into a run, trips over a leaf stem, and somersaults ungracefully into a hollow. Shaken, the salamander rights itself and climbs out of the depression, finally ducking its head to slide under a dead leaf. Cold mist thickens the air, and I can see only a few feet ahead of me, but the salamander shines as if it were illuminated by a clear ray of sunlight. The dark, smooth skin is freckled with silver. Small red streaks flow down the animal's back. The skin is impossibly wet, a cloud condensed into animate matter.

Like mosses, salamanders thrive on moisture, but salamanders cannot use the mosses' strategy of drying up and waiting out the days between rains. Instead, they follow cool, humid air like nomads, moving in and out of the soil as the humidity changes. In winter they creep down between rocks and boulders, escaping the freeze and living as troglodytes in the subterranean darkness, up to seven meters belowground. In the spring and autumn they climb back up and ply the leaf litter, pursuing ants, termites, and small flies. Summer's drying heat pushes them back underground, although on wet summer nights salamanders burrow back to the surface to feast without danger of dehydration.

The salamander is twice as long as my thumbnail. Its neck and legs are slender, marking it as a member of the genus *Plethodon*, perhaps a zigzag salamander or a southern redback. The fact that all *Plethodon* species are variably colored and poorly studied reinforces the imprecision of my identification. Then again, no one is quite sure what a salamander "species" really is, which suggests that nature doesn't conform to our desire to draw firm lines.

The salamander is small, so it is likely a juvenile, hatched late last summer. Its parents courted last spring, with delicate footwork and tender cheek rubbing. Salamander skin is a patchwork of scent glands, so the cheek rubs convey chemical whispers and pheromone love poems. When the couple has become acquainted, the female lifts her head and the male slides under her chest. He walks forward and she follows, straddling his tail in a conga dance for two. After a few steps, he deposits a small cone of jelly topped with a packet of sperm. He moves forward again, waggling his tail, and the female follows. She stops and uses her muscular vent to pick up the sperm. The dance breaks up and the salamanders wander on their separate ways, never to interact again.

The female seeks out a rock crevice or hollow log in which to lay her eggs. She then wraps herself around them, remaining in the nest hole for six weeks, longer than most songbirds sit on their eggs. She rotates the eggs to stop the developing embryos from sticking to the sides. She also eats any egg that dies, preventing mold from growing and killing the whole clutch. Other salamanders may visit the nest hole, looking for an egg snack, and the brooding mother chases them off. Motherless broods invariably get infected by fungi or eaten by predators, so this vigil is crucial. Once the eggs hatch, her parental duties are finished, and the mother will renew her depleted energy reserves by feeding in the leaf litter. The young salamanders are miniature versions of the parent and strut across the forest floor, feeding themselves without assistance. The *Plethodon* scuttling across the mandala therefore lives its whole life without dipping a toe into a stream, puddle, or pond.

This breeding process demolishes two myths. The first is that amphibians are dependent on water for breeding—*Plethodon* is a nonamphibious amphibian, as slippery to classify as it is to hold. The second myth is that amphibians are "primitive" and therefore don't care for their young. This latter fallacy is embedded in theories about the evolution of the brain claiming that "higher" functions such as parental care are confined to "higher" animals such as mammals and birds. The mother's careful vigil shows that parental solicitude is more widely spread in the animal kingdom than hierarchical brain scientists suppose. Indeed, many amphibians care for their eggs or their young, as do fish, reptiles, bees, beetles, and a menagerie of doting "primitive" parents.

The juvenile salamander in the mandala will spend another year or two feeding in the leaf litter before it is large enough to become sexually mature. *Plethodon* sets to this task of feeding with carnivorous gusto. Salamanders are the sharks of the leaf litter, cruising the waters and devouring smaller invertebrate animals. Evolution has discarded *Plethodon*'s lungs to make its mouth a more effective snare. By eliminating the windpipe and breathing through its skin, the salamander frees its maw to wrestle prey without pause for breath. *Plethodon* has struck a deal with evolution's Shylock: better tongues bought with a few grams of lung. The salamanders are living it up on their three-thousand-ducat loan, conquering the wet leaf litter across the eastern forest. The gamble is paying off at present, but the usurer may yet call in the debt. If pollution or global warming changes conditions in the leaf litter, *Plethodon* species will be ill suited to cope. Indeed, projections of habitat change caused by global warming suggest that mountain salamanders will suffer major declines as their cool, wet habitats disappear.

No one knows how *Plethodon* salamanders arrived at their lungless condition. Their relatives all have lungs, although those that live in mountain streams have rather small lungs. Cold streams have plentiful oxygen, so stream-dwelling salamanders can use their skin as a breathing organ. Perhaps the terrestrial lungless salamanders evolved from these stream-dwelling kin? This was biologists' favorite explanation until researchers looked more closely into the geological record. The rocks told an inconvenient story: the eastern mountains were small undulations when *Plethodon* salamanders evolved. Such gentle inclines could not have produced the cold, rapid streams inhabited by the small-lunged salamanders. So, we are left without a historical narrative for the *Plethodon*'s lungless condition.

The mandala is almost large enough to contain the whole world of this animal. Adults are territorial and rarely stray more than a few meters; some individuals move farther downward into the soil than they do across the surface of the litter. This rootedness accounts for the diversity of woodland salamanders. Because they seldom move far, the salamanders on different sides of a mountain or valley are unlikely to interbreed. Local populations therefore adapt to the particularities of their habitat. If this divergence keeps up for long enough, separate populations may come to look different and have different genetic characteristics. Some may even get called different "species," depending on the current taxonomic fashion. The Appalachian Mountains are ancient rocks, and their southern end, where the mandala sits, has never been covered by a killing sheet of ice-age glaciers. The salamanders here have therefore had time to explode in a burst of diversity that is unmatched anywhere on the planet. This diversity partly accounts for why salamanders are so difficult to classify into species.

Unfortunately for the salamanders, the old wet, warm forests that produced salamander diversity also grow large, profitable trees. If these trees are removed in large clear-cuts, the shady leaf litter turns into a sun-beaten crisp, killing all the salamanders. If the clear-cut is lucky enough to be surrounded by mature forest, and if it is left alone for several decades, salamanders will slowly return. But the salamanders do not return to their former abundance, although no one knows why. Perhaps large clear-cuts eliminate genetic fine-tuning from local populations? Logging also removes trees that would have fallen to create moist crevices, nesting holes, and refuges from the sun. The scientific jargon for these life-giving fallen trees is "coarse woody debris," a term that seems too dismissive for such a life-giving part of the forest's ecology.

The salamander in the mandala thrives among the messy tree falls in this small protected patch of old-growth forest, but although clearcuts are unlikely, the animal is not free from danger. This salamander is tailless, probably the result of an encounter with a mouse, bird, or ringneck snake. When attacked, salamanders thrash their tails to divert the predator. If needed, the tail will break off and undulate violently, providing distraction while the salamander escapes. The blood vessels and muscles at the base of *Plethodon* tails are specially adapted to clamp shut once the tail is lost. The base of the tail also has weaker skin and is constricted, presumably to help the tail break free without hurting the rest of the body. Evolution has therefore struck two bargains with these animals, both secured with flesh: better mouths bought with lunglessness and longer lives bought with detachable tails. The first deal is irreversible; the second is temporary, erased by the mysterious regenerative power of the tail.

Plethodon is a shape-shifter, truly a cloud. Its courtship and parenting defy our haughty categories, its lungs were traded for stronger jaws, its body parts are detachable, and it is paradoxically moisture loving yet never enters bodies of water. And, like all clouds, it is vulnerable to strong winds.

March 13th—*Hepatica*

The temperature has been warm all week, giving us an unseasonable but welcome foretaste of May. The first spring wildflowers have sensed the change and have pressed up from below the litter, causing the formerly smooth mat of dead leaves to buckle as the stems and buds of flowers elbow through.

I shed my shoes for the first part of my walk to the mandala, treading barefoot on the worn public foot trail, feeling the ground's mild warmth. Winter's sharpness is gone. As I walk in the gray predawn light, birds are in full song. Phoebes rasp from the rocky bluff, accompanied by titmice whistling from low branches and woodpeckers cackling from large trees below the trail. Aboveground and below, the season has turned.

At the mandala, I find that one flower bud, a *Hepatica*, has finally pierced the litter, standing on a finger-high stem. A week ago, the bud was a thin claw, encased in silver fuzz. Slowly, the claw filled out, fattening and elongating as the air warmed. This morning, the bud's stem is shaped like an elegant question mark, still covered in down, with the tightly closed flower suspended at the tip of the curve. The flower points demurely down, its sepals closed against nighttime raiders of pollen.

The flower cracks open an hour after first light. The three sepals spread, revealing the edges of three more inside. The sepals are flushed with purple and, although *Hepatica* lacks true petals, these sepals have the shape and function of petals, protecting the flower at night and attracting insects by day. The flower's opening motion is too slow for my eyes to perceive directly. Only by looking away then returning my gaze can I see the change. I try to still my breathing, slowing to flower-speed, but my brain races too fast, and the slow, graceful motion eludes me.

Another hour passes and the stem straightens; the question mark turns into an exclamation point. The sepals are spread wide now, shining rich purple at the world, inviting bees to investigate the untidy mop of anthers at their center. One more hour and the exclamation point is written in a hurried hand, bent backward a little, lifting the flower's face directly at me. This is the mandala's first bloom of the year. The lively skyward arch of the flower's stem seems a fitting gesture of springtime release and celebration.

The flower's name, *Hepatica*, has a long history, one that reaches back to Western Europe where a close relative of the same name has been used in herbal medicine for at least two thousand years. Both the scientific name and the common name, liverleaf, refer to the plant's purported medicinal qualities, suggested by the three-lobed liverlike shape of the leaves.

Most of the world's cultures have a habit of extrapolating from the shapes of plants to their medicinal powers and hence to their names. In the Western tradition, this habit was codified into a theological system by an unlikely scholar. In 1600, a German cobbler, Jakob Böhme, experienced a stunning vision of God's relationship to creation. The heart-ripping magnitude and power of this revelation tore him away from his shoe-making trade and thrust a quill into his hand. Out flowed a book, a stream of words attempting to communicate the massive wordless vision. Böhme believed that God's purpose for His creation was signed into the forms of worldly things. Metaphysics was scrawled into flesh. He wrote, "Every thing is marked externally with that which it is internally and essentially . . . [and] represents that for what it may be useful and good." Mortal, imperfect humans could therefore deduce purpose from the outward appearance of the world and could see the thoughts of the creator in the shapes, colors, and habits of His creation.

Böhme's work caused his expulsion from his hometown, Görlitz. The church and the city council would not tolerate unauthorized mystical experiences. Shoemakers, they felt, should stick to cutting leather and leave visions to the well-read and the well-bred. Later, he was allowed to return on the condition that he keep his quill away from paper. He tried and failed, the vision's power pushing him on to Prague, where he continued his theological essays.

Böhme's ideas did not become widely known until botanical physicians learned of his work. His doctrine helped their trade by providing a theological cabinet in which to store their herbal remedies. Many physicians already used the external forms of plants as mnemonics with which to remember their medicinal functions: the scarlet juice of bloodroot for disorders of the blood, toothwort's indented leaves and white petals for toothache, coiled roots of snakeroot for snakebite, and dozens more. Now, the healers had a theory with which to organize and justify their practices. The shape, color, and growth of plants indicated their divine healing purpose. The showy, scented blossoms of the apple tree were intended to heal disorders of fertility and complexion; red, peppery plants were stamped with the sign of blood and anger, and so could be used to stimulate the circulation or the spirit. The *Hepatica*'s three-lobed purple leaves bore the liver's mark.

The use of external marks to deduce and remember the medicinal function of chemicals inside plants became known as the Doctrine of Signatures. The idea spread across Europe and attracted the attention of the scientific elite. They tried to haul the herbalists' doctrine out of folklore and into the then modern science of astrology. The signature in each plant, they claimed, reflected God's purpose, but it did so through the complex cosmology of the planets, moon, and sun. The apple blossom was governed by Venus, hence its beauty and its healing powers. Jupiter governed all hepatic plants, and Mars ruled the warlike peppers. Correct diagnosis and treatment therefore required that a qualified scientist cast a horoscope and create a remedy incorporating his extensive, expensive knowledge of the celestial spheres and their influence on both plants and the human body. The scientific establishment railed against the country quackery of the simpleminded botanists while expropriating the quacks' remedies for use in an updated astrological medicine.

This tension between the medical establishment and the quacks continues, of course. The astrological Doctrine of Signatures now finds itself out of favor. Our physicians no longer believe that God left providential medicinal hints in the shapes of leaves and in the arrangement of the stars. We should not, however, be too quick to dismiss the Doctrine of Signatures as a trifling superstition. As a method of cultural transmission of medical knowledge, the doctrine was a powerful organizing device, richer and perhaps more coherent than the mnemonics used by modern physicians to navigate their large stores of knowledge. The method gave healers, most of whom could not read, linguistic cues to connect patients' symptoms with the sometimes arcane details of botanical identification and medical knowledge. The Doctrine of Signatures persisted for so many years not because our ancestors were simpleminded but because it was so useful.

Hepatica's name reveals our culture's propensity for naming plants after their uses. This method of naming helps us to remember humanity's dependence on plants for medicines and foods. But utilitarian names can also stand in the way of a full experience of nature. For example, our nomenclature has its teleology wrong. *Hepatica* exists not to serve us but to live out its own story, one that began in the forests of Europe and North America millions of years before humans came to be. Likewise, our naming imposes tidy categories on nature. These may not reflect life's complicated genealogies and reproductive exchanges. Modern genetics suggests that boundaries in nature are often more permeable than we suppose when we name "separate" species.

On this bright morning in early spring, the *Hepatica*'s confident welcome of the first warm sun and flying bees reminds me that the mandala exists independent of human doctrines. Like all people, I am culture-bound, so I only partly see the flower; the rest of my field of vision is occupied by centuries of human words.

March 13th—Snails

The mandala is a molluskan Serengeti. Herds of coiled grazers move across the open savanna of lichens and mosses. The largest snails travel alone, plying the crazy angled surfaces of the leaf litter, leaving the mossy hillsides for the nimble youngsters. I lie down on my belly and creep up on a large snail that sits at the edge of the mandala. I lift the hand lens to my eye and shuffle closer.

Through the lens, the snail's head fills my field of vision—a magnificent sculpture of black glass. Patches of silver decorate the shining skin, and small grooves run across and down the animal's back. My movements cause mild alarm; the snail withdraws its tentacles and hunches back into the shell. I hold my breath and the snail relaxes. Two small whiskers poke their way out of the chin, waving in the air before reaching down and touching the rock. These rubbery feelers move like fingers reading Braille, touching lightly, skimming meaning from the sandstone script. Several minutes later a second pair of tentacles launches out from the crown of the snail's head. They reach upward, each with a milky eye at its tip, and wave at the tree canopy above. My own eye bulges at the snail through the lens, but this monstrous globe seems to be of little concern to the snail, which extends its eyestalks farther. These fleshy flagpoles now reach wider than the shell and swing wildly from side to side.

Unlike its relatives the octopi and squid, this land snail has no sophisticated lens and pinhole through which to form crisp images.

But just how fuzzy the world appears to a snail is a mystery. Scientists have difficulty asking the snails what they perceive, and this communication problem slows the leading edge of snail vision research. The only experimental success in this area has come from borrowing the tricks of circus trainers and teaching snails to eat or move when they see a signal. So far, these performing gastropods have shown that they can detect small black dots on a white test card. They can also distinguish between gray and checkered cards. As far as I know, no one has yet asked a land snail whether it can see color, motion, or a flaming hoop.

These experiments are fascinating, but they leave aside a larger question: what is a snail "seeing"? Do snails see as we do, with images of checkered cards appearing in their gastropod minds? Do they experience private displays of light and dark, processed by tangles of nerves into decisions, preferences, and meaning? The human body and the snail body are made from the same wet pieces of carbon and clay, so if consciousness grows out of this neurological soil, on what grounds do we deny the snail its mental images? No doubt what it sees is radically different, an avant-garde movie of strange camera angles and lurching forms, but if the human cinema is caused by nerves, we have to allow for the startling possibility that the snails have a similar experience. But our culture's preferred story is that the snail movie plays to an empty house. Indeed, the theater has no screen. The snail has no internal subjective experience, we claim. Light from the eye's projector merely stimulates the snail's ductwork and wiring, causing the hollow theater to move, eat, mate, and keep up the appearance of life.

The snail's head explodes, ending my speculations. The black dome is split by a knot of cloudy flesh. The knot pushes out, forward, then the snail turns to face me. The tentacles form an X, radiating away from the bubbling, doughy protrusion at the center. Two glassy lips push out, defining a vertical slit, and the whole apparatus heaves downward, pressing the lips to the ground. I watch, saucer-eyed, as the snail starts to glide over the rock, levitating across a sea of lichen. Tiny beating hairs and ripples of infinitesimally small muscles propel the ebony grazer on its path.

From my prone position I see the snail pause amid lichen flakes and black fungus spiking from the surface of oak leaves. I peek over the lens and suddenly it is all gone. The change of scale is a wrench into a different world; the fungus is invisible, the snail is a valueless detail in a world dominated by bigger things. I return to the lens world and rediscover the vivid tentacles, the snail's black-and-silver grace. The hand lens helps me harvest the world's beauty, throwing my eyes wide open. Layers of delight are hidden by the limitations of everyday human vision.

My snail vigil ends when the sun breaks out from behind a cloud. The morning's soft humidity has lifted, and the snail heads toward El Capitan, or a smallish rock, depending on how you see the world. Here the snail touches a tentacle to the rock, then turns its entire head upside down and stretches up. The neck and head rubber-band into a giraffe's, farther, a little farther, then the chin hits the rock, spreads itself into a pad, and the whole snail lifts up from the ground in a no-handed chin-up. Gravity blinks and the animal flows impossibly upward and continues its journey, upside down, into the rock crevice. I look up, out of the lens world, and the Serengeti has emptied. The grazers have evaporated in the sun.

March 25th—Spring Ephemerals

My walk to the mandala has become fraught. Every footstep threatens to squash half a dozen wildflowers, and so I step slowly, trying to pick out a way that does not leave a trail of crushed beauty. The mountainside is heavily peppered with green and white; half the leaf litter's surface is covered by newly grown leaves and flowers.

It is hard, though, to concentrate on my feet when the year's first butterflies and migratory warblers are flying above me. An eastern comma, a rufous butterfly named for the white curl on its hindwing, flicks past my head and lands on a hickory trunk. The warm sun has roused it from its winter hibernation, hidden under a bark flake. A black-throated green warbler and a black-and-white warbler, both recently returned from Central America, sing from the bluff. The forest's renewed life seems to crowd in on me from all sides, lifting my spirits with its unrestrained vigor.

At the mandala I find a starburst of white flowers, a hundred blossoms shining out at the world. Spring beauty flowers with pink-streaked white petals grow low to the ground, intermixed with purple *Hepatica*. A few rue anemones emerge from the mandala's edge, their nodding white flowers held finger-length above the leaf litter. Toothwort reaches tallest, just above ankle height, holding flowers with long white petals in clusters at the tips of sturdy stalks. Each flower trails a comet's tail of lush green growth, erupting life from the mat of dead leaves. The contrast with the wintry trees above the mandala is dramatic. Tree buds are barely broken open.

Spring wildflowers take advantage of the trees' sluggishness, rushing through their reproduction and growth before the tree canopy steals life-giving photons. Although the March sun is still low, its rays are strong enough to burn the back of my neck as I sit. We have reached the peak of the year's cycle of light intensity below the canopy. Winter's hold is broken with blazing force, unlocking constellations of flowers and a cascade of animal life.

The plants that festoon the mandala are collectively called spring ephemerals. This name captures their meteoric brilliance in springtime and their rapid fade in the summer sun, but the name belies their secret underground longevity. These plants grow from subterranean storehouses, some of them growing from hidden belowground stems called rhizomes, others emerging from bulbs or tubers. Every year, the plants send up leaves and flowers, then return to covert quiescence. The flowers' push into the cool spring air is therefore fueled by food stored up from the previous year. Only after the plants have leafed out does photosynthesis boost their balance sheet. This strategy helps them persist in the choked, light-hungry world of the mandala. Some of these stems may be hundreds of years old, having slowly crept across the forest floor by growing a few centimeters of horizontal stem each year. These plants survive on the food gained during a few short weeks of springtime sun.

Once the ephemerals have unfurled their leaves, they reap sunlight and carbon dioxide at a furious rate. The breathing holes in their leaves, the stomata, are thrown wide open. Leaves are stuffed with enzymes ready to concoct nutritious molecules out of air. These plants are the fast-food junkies of the forest: they eat rapidly, rushing to get through the meal before the trees block out the light. The ephemerals require bright sunlight to sustain this gluttony. Their hyped-up bodies cannot tolerate shade.

Other plants in the mandala take a slower road. The toadshade trillium pokes up a trio of dappled leaves between the *Hepatica* and spring beauty plants, but it is not seeking a quick spurt of growth. Toadshade trillium's leaves have few enzymes with which to harness sunlight, so they cannot match the ephemerals' growth rate. Their thrift is rewarded when the tree canopy closes; low levels of enzymes are cheap to maintain, so the trillium can make a sugary profit in the deep shade of summer. We are at the starting line for the annual botanical race for the mandala's limited space. Evolution has produced a wonderful diversity of running styles: Carolina spring beauty is a muscled-up sprinter, trillium is a lean distance runner.

The bright burning lives of the ephemerals ignite the rest of the forest. Their growing roots reinvigorate the dark life of the soil, absorbing and holding the nutrients that would otherwise be flushed out of the forest by the spring rains. Each root secretes a nutritious gel, creating a sheath of life around its hairy tip. Bacteria, fungi, and protists are a hundred times more abundant in this narrow halo, and these single-celled creatures provide food for nematodes, mites, and microscopic insects. The grazers are preyed upon by even larger soil-dwellers such as the bright orange centipede that shimmers back and forth over the mandala as I sit watching. The centipede is longer than my hand is wide, so big that I can see each segment on its legs as the body undulates between the sources of its life, the flowers.

A few days ago my contemplation of the flowers was broken by a fiercer predator than this centipede. A palm-sized ball of gray fur shot out of the ground, then dove back into another hole, accelerating like a dustball pulled into a vacuum cleaner. A few minutes later I heard rustling and high-pitched squeaks from the other side of the mandala. I saw just enough of the sooty fur and stubby tail to know that the terror of the leaf litter was prowling the mandala: a short-tailed shrew.

Shrews live short, violent lives. Only one in ten survives longer than a year; the rest get burned out by their furious metabolism. Shrews breathe so frantically that they cannot survive long aboveground. Their outrageously rapid breathing would desiccate and kill them in the dry air.

Shrews feed by snapping at prey then chewing poisonous saliva into their victims, sometimes killing the animal they have caught, sometimes paralyzing it for storage in a dungeon of horrors, a larder of living but incapacitated prey. So ferocious are shrews that they eat whatever is before them. Mammalogists despair of them. If a shrew gets caught with mice in a live trap, the scientists return to find a pile of bones overseen by an angry gray warden.

The squealing that I heard was just the lower range of the shrews' repertoire. Most of their chattering is too high for my ears to detect. These highest calls are shrew sonar. Shrews send out ultrasonic clicks and listen for the reflected sound waves, using echolocation to find their way around burrows and to locate prey. These "terrestrial submarines" therefore navigate mostly by sound. Their eyes are tiny, and mammalogists disagree about whether shrews can see images or whether they merely perceive patches of light and dark. As with the snails, shrew vision is a mystery.

The soil's food web reaches its zenith in the shrew. Only owls will eat shrews; everything else gives them a wide berth, fearing their vicious teeth or the acrid taste of their scent glands.

There is kinship with humans here. The first mammals were shrewlike creatures terrorizing the snails and centipedes of the Mesozoic. Our ancestors were shrill and vicious, leading a caffeinated existence in dark corridors. An analogy with our current state of being is tempting. Thankfully we've lost the poison fangs and pungent glands.

The spring ephemerals also touch off life's fire aboveground. Small black bees fly from flower to flower, rejecting all but Carolina spring beauty flowers. Here the bees bury their heads, slaking their thirst for the strong sugar water that we call nectar, then whir their legs through the flower's pink pollen-bearing anthers. The bees emerge looking like chocolate drops dusted with rosy icing sugar. They fly off with fat saddlebags of pink pollen hanging from each hind leg.

The flying candies are all female bees, recently emerged from their winter burrows. Each female flies around seeking a new nest site in either a patch of soft soil or an old log. The bees dig tunnels into their chosen home and paint shiny secretions on the walls of a nest chamber. These secretions hold the walls together and keep water away from the delicate offspring. The mother mixes pollen and nectar into a ball, then lays an egg on the ball and seals it into a small mud-walled cell. The larval bee that hatches from the egg will munch its way through the pollen paste and emerge several weeks later, its body built wholly out of flowers. This dependence on pollen and nectar will continue for the rest of the bee's life. Bees eat nothing else; they are the original "flower power" creatures.

Once emerged, the offspring of some species of woodland bee fly off to breed on their own. Many other species stay at home, forgoing the opportunity to lay their own eggs. These helper bees take over the foraging duties, allowing the foundress, their mother, to specialize in egg laying. This communality is favored by two forces, one external to the bees and one built into their genes.

The crowdedness of the bees' environment encourages stay-at-home offspring. Most of the forest floor is too rocky, too wet, or too deeply covered in leaf litter to make an adequate nest. Competition for nest sites is intense, and female bees that try to strike off on their own face a serious risk of failure. Staying at home is a safer bet; if you've been born there, then by definition your mother has a successful nest hole.

The bees' genetics tip the scale further in favor of helping mother. Female bees are born from eggs fertilized by sperm stored from the mother's autumnal love flight, so they have, as do humans, two copies of all their chromosomes, one from mom and one from dad. In contrast, males develop from unfertilized eggs, so they carry only one set of chromosomes, inherited only from their mother. Therefore all bee sperm cells are identical. This curious genetic system produces even more curious kinships. Sisters within a bee colony are very closely bonded, a supersorority of chromosomes. Whereas human siblings share, on average, half their genetic individuality, these bee sisters share much more. The half of their DNA that they inherit from their father is identical; the half they inherit from their mother is split evenly among sisters. The average of their parents' contributions therefore comes to three-quarters of their genes via common descent. If the bee mother mated with more than one male, this relatedness drops a little but still remains high enough to affect the evolutionary process.

Evolution's accountant rewards those animals that assist close kin and ignore more distant relatives. Normally this means that raising one's own offspring is the best strategy. But female bees' genes have primed them to be as open to helping mom as they would be to leaving home and breeding themselves. So, when the mother bee fills her springtime nest with fertilized eggs, she is hatching a cohort of daughters for whom leaving home is risky and staying at home is very attractive. Male bees feel a different set of forces. No strange relatedness rewards them for staying at home. Sons therefore behave like aristocratic cads, hanging about the nest, ambling about in search of nectar, and focusing their energies on the pursuit of virgin queens. Their sisters have little patience with them and will sometimes forcibly expel them from the nest.

Brother-sister tensions are not the only source of conflict in the bee nest. Worker bees sometimes try to sneak their own eggs into the nursery. The queen responds by eating the eggs and releasing odors that suppress egg laying in her uppity daughters, reinforcing the already strong pull of genetic relatedness. Sometimes several overwintering females will found a colony together, causing a tug-of-war over who lays the most eggs. The winner usually becomes queen, but her cofoundresses continue to try to lay eggs themselves.

Fraught family lives are not the only source of woe in bee nests.

Defenseless larvae and concentrated stores of pollen and honey make alluring targets for raiders. Many of these raiders are out in force today over the mandala's flowers. Bombyliid flies or "bee flies" are the most specialized and successful of these pirates. Adult bombyliid flies are innocuous, comical even. They dart from flower to flower, supping on nectar with a rigid proboscis that points the way for the orange feather-duster-fluffy body. But this cuddly flower-tippling buffoonery ends when the female drops her egg in front of a bee nest. The egg hatches, and a wormy larva crawls into the nest to dine on the bees' pollen and honey stores. The worm then molts into a predaceous larva that consumes the bee larva whose cell has been robbed. Sated, the fly larva encases itself and waits underground. When the ephemerals kick-start the mandala's life next spring, the bombyliid flies crawl out of their pupal dens, metamorphosed from pirates into clowns.

A pattern emerges as I watch the bees and flies in the mandala. Adult bombyliid flies show no discernment in their choice of flowers, stopping at every flower to sip nectar or eat pollen. The bees are fussier, preferring spring beauty and rejecting the nectarless flowers of rue anemone and *Hepatica*. These preferences are the hem of a huge and complex cloak of relationships. Hundreds of species of insects and flowers interact in this forest every spring, each trying to buy success for its offspring through sugary bribery or subsidized hauling of pollen. Some, like the bombyliid flies, are numerous but only partly successful at transferring pollen. Others, like the fussy bees, are rarer but more effective pollinators.

This intricate web of dependency dates back one hundred and twenty-five million years to when the first flowers evolved. The oldest fossil flower, called *Archaefructus*, had no petals, but its pollen-bearing anthers had flags on their tips. The botanists who described the fossil believe that these extensions may have been used to attract pollinators. Other ancient flowers also appear to have been insect-pollinated, further supporting the idea that insects and flowers have been partners since the first flowers evolved. How this marriage came about is unknown, but it seems likely that flowering plants evolved from fernlike plants. These ancestors produced spores that attracted insects looking for an easy meal. The ancestors of the flowers turned the plague of insect predators into a blessing by producing conspicuous displays to attract these spore munchers, then producing so many spores that the insects' bodies would be coated. The predators inadvertently carried some of this sporey dust onto the next flower, increasing the fecundity of the spore producer. Eventually the spores got wrapped in a package, the pollen grain, and the true flower was born. The bees and spring beauties in the mandala reenact the main theme of the original relationship. The bees, or their larvae, eat most of the pollen they gather, transferring only a small number of pollen grains from flower to flower.

The core of the relationship between flowers and insects is unchanged, but the details and frills have been elaborated enormously. An insect flying across the mandala is bombarded with scents, billboards, and lures, all trying to coax her to the flowers' storefronts. The bombyliid flies answer all these calls, stopping in at every flower. Most bees are more selective. Sometimes this selectivity produces specialization: a flower designed for one insect, an insect's brain tuned to one flower. The orchids carry this to exquisite extreme, mimicking the aroma and appearance of a female bee, inducing mating by the male, whose ardor is then converted into an orchid postal system.

The mandala has a small number of specialized flowers. Toothwort's tubular flowers exclude small bees, allowing only long-tongued bees and flies into the narrow nectar tube. Some bee species feed exclusively on spring beauty flowers, choosing to be faithful to one flower for the sake of efficiency. But these examples of specialization are conspicuous exceptions to the promiscuity of the plants and their pollinators in the mandala. Spring's brevity promotes this unusual preponderance of generalists. The ephemerals are caught between the cold weather of early spring, which stops pollinators from flying, and the tree canopy's emergence, which steals the light that the spring ephemerals need to grow and produce seeds. This is no time for pickiness. Plants need any help they can get from the insects, regardless of whether the insect is a faithful bee or a haphazard fly. All the flowers in the mandala, except for the toothwort, produce cup-shaped flowers that are available to any insect. The starburst is open wide, welcoming all the woodland pollinators to its blazing show.

April 2nd—Chainsaw

A mechanical whine starts abruptly and cuts through the forest, jangling my nerves as I sit at the mandala. A chainsaw is ripping through wood somewhere to the east. This patch of old-growth forest is protected, supposedly free of chainsaws, so I leave the mandala to investigate. After scrambling across a rock scree and climbing up a stream bank, I find the source of the sound: a golf course maintenance crew felling a dead tree at the edge of one of the cliffs above the forest. The golf course runs to the edge of the cliff, and dead trees evidently do not fit with the golfing aesthetic. The maintenance crew bulldozes the felled tree over the edge of the cliff, then moves on to other tasks.

The sight of a cliff being used as a disposal chute is irritating, but the dumped tree will provide some extra salamander habitat. I am relieved that the cutting was not below the cliff line in the old-growth forest itself. The mandala's intense starburst of flowers is special, and nearly unique, because the chainsaw has never stripped this hillside of its cover. Salamanders, fungi, and solitary bees also revel in the tangles of huge fallen logs and deep leaf litter. Logging, particularly clear-cutting, kills many of these inhabitants of the woods, and their populations take decades, sometimes hundreds of years, to recover.

Stripping the mountainside of trees turns the forest soil from a moist duff to an oven-baked brick. Ground-nesting bees, wet-backed salamanders, and the creeping stems of ephemerals dry up and die in the parched soil. Only when the forest has regained its leaf litter, canopy, and dead wood do the creatures start to return, but this revival is slow, constrained by the lack of old dead logs to act as nurseries and by the sluggish dispersal abilities of flowers and salamanders.

So what? Why should we leash our accelerating desire for wood and paper for the sake of saving a springtime explosion of forest biodiversity? Can't the flowers look after themselves? After all, disturbance is natural. The old "balance of nature" cliché went out of fashion decades ago. Now the forest is a "dynamic system," constantly assaulted by wind, fire, and humans, always in motion. Indeed, we can turn the question on its head and ask whether we *need* to go out and do some clear-cutting, to replace the fires that used to clear large areas of forest but have been suppressed by land managers for nearly a hundred years now.

These questions are the roots of a thorny growth of arguments that choke academic conferences, government reports, and newspaper editorials. Does the forest need the buzz of chainsaws, or does it need time to renew itself unperturbed by log hunters? We are tempted to use nature as a model, but nature provides a Baskin-Robbins of justifications. Which flavor of forest life cycle would you like: the annihilating force of an ice age; or an ancient, undisturbed mountainside; or the dancing mischief of a summer tornado?

Nature, as usual, is not providing the answer.

Rather, we are thrown back a moral question: what part of nature do we wish to emulate? Do we aspire to the uncompromising, all-controlling weight of an ice sheet, imposing our glacial beauty on the land, retreating every hundredth millennium to free the forests' slow regeneration? Or do we seek to live like fire and wind, cutting swaths with our machines, then moving on for a while, hitting at random intervals, at random locations? How much wood do we need? How much do we desire? These are questions of time and of magnitude. We can cut every two decades, or we can cut every two centuries; we can focus our extractive desires, or we can let them run across the whole; we can strip the forest bare, or we can remove just a few trees.

Our collective answer to this question grows out of the values of millions of landowners, pruned and guided by society's two clumsy hands, the economy and government policy. The forest is shattered like a broken windshield by surveyors' lines, so the diversity of these values plays out in a mosaic across the continent. Despite this chaos, patterns emerge from the aggregate. We are neither an ice age nor a windstorm but something altogether new. We have changed the forest on the scale of an ice age, but at a pace accelerated a thousandfold.

In the nineteenth century we stripped more trees from the land than the ice age accomplished in one hundred thousand years. We hacked the forest down with axes and handsaws, hauling it away on mules and railcars. The forest that grew back from this stripping was diminished, robbed of some of its diversity by the scale of the disturbance. This was a windstorm on the scale of an ice age but similar to a tornado in its crude physical messiness.

Cheap oil and expensive technology have now brought us to the second phase of our relationship with the forest. No longer do we cut by hand and haul with animals or steam; gasoline engines do it all, accelerating our extraction and increasing our control. Oil's power and our minds' cleverness gave us another tool: herbicides. In the past, the forest's regenerative power limited our ability to direct the land's future. The forest would burst back, prepared for the ax by millions of years of wind and fire. Now "chemical suppression" is the tool of choice to knock back trees whose genes tell them to resprout. Machines clear the forest, cutting then bulldozing the remaining "debris." Then the helicopters move in and rain herbicides on the remnants, forestalling a resurgence of green. I have stood at the center of these clear-cuts and seen no green to the horizon in nearly every direction: an arresting experience in Tennessee's normally lush summer.

All this effort is directed at preparing the land to receive a new forest, a monoculture of fast-growing trees. Depending on the tree and the soil type, the rows of trees are then sprayed with fertilizers to replace some of the nutrients that were removed from the old, outmoded forest. If you squint, the resulting tree plantations look something like forests. But the diversity of birds, wildflowers, and trees is gone. Suburban backyards have more biological diversity than these shadows of real forests.

Can plantations ever be restored to foresthood? The lesson of the ice age is that such annihilation can be reversed, but at a pace measured in millennia, not decades. But the question is premature. The ice is not pulling back. Every major type of native forest in the southeastern United States is in decline. Only plantations are on the rise.

The scale, the novelty, and the intensity of this change are unquestionable threats to the diversity of life in the forests. Whether or how we should respond to this erosion is a moral question. Nature seemingly provides no moral guidance; mass extinction is another of her many flavors. Neither can moral questions be answered by our culture's obsession with policy think tanks, scientific reports, or legal contests. I believe that the answers, or their beginnings, are found in our quiet windows on the whole. Only by examining the fabric that holds and sustains us can we see our place and, therefore, our responsibilities. A direct experience of the forest gives us the humility to put our life and desires into that bigger context that inspires all the great moral traditions.

Can the flowers and bees answer my questions? Not directly, but two intuitions come to mind by contemplation of a multifarious forest whose existence transcends my own. First, to unravel life's cloth is to scorn a gift. Worse, it is to destroy a gift that even hardheaded science tells us is immeasurably valuable. We discard the gift in favor of a self-created world that we know is incoherent and cannot be sustained. Second, the attempt to turn a forest into an industrial process is improvident, profoundly so. Even the apologists for the chemical ice age will admit that we are running down nature's capital, mining the soil, then discarding the spent land. This rash ingratitude, justified by the economic "necessity" created by our ballooning consumption of inexpensive wood, seems to be an outward mark of inner arrogance and confusion.

Wood and wood products such as paper are not the problem. Wood provides us with shelter, paper with nourishment for the mind and spirit—unarguably wholesome outcomes. Wood products can also be much more sustainable than the alternatives such as steel, computers, and plastic, all of which use large quantities of energy and nonrenewable natural products. The problem with our modern forest economy lies in the unbalanced way that we extract wood from the land. Our laws and economic rules place short-term extractive gain over all other values. It does not have to be this way. We can find our way back to thoughtful management for the long-term well-being of both humans and forests. But finding this way will require some quiet and humility. Oases of contemplation can call us out of disorder, restoring a semblance of clarity to our moral vision.

April 2nd—Flowers

An impossible number of flowers blaze out of the mandala. Confusion sets in when I try to count them: two hundred and eighty, three hundred and twenty, too many crowded into one square meter. The flowers' valets are in attendance, buzzing and humming in smart furry dress, fussing over the floral royalty. I join them in their observance and genuflect, then prostrate myself, hand lens pressed to my eye.

A fountain of anthers arches from the chickweed's open bloom. A central dome, the ovary, is ringed by gracile, creamy filaments holding up tawny knots of pollen grains. These filaments soar away from the dome, holding the pollen away from the flower's own pollen landing pads, the stigmas. The chickweed has three stigmas, planted at the peak of the ovary's onion dome, each one waiting for a pollen-dusted bee to brush past.

The surface of the stigma is a forest of microscopic fingers, reaching out to embrace pollen grains. If the petals do their job and attract a bee, the stickiness of the stigma traps the rough-coated grains. Once pollen is caught, the stigma assesses it, rejecting any from different species. The plant also shuns its own pollen and that from close relatives, preventing self-fertilization and inbreeding. In a few species, this rule against self-fertilization is broken if no other suitable pollen arrives. Such self-fertilization is a strategy of last resort used by *Hepatica* and other species that bloom in the early spring. For these species, desperate self-love is better than no love at all when inclement weather grounds their pollinating insects.

If the biochemical matchmaking goes well, the stigma's cells release water and nutrients to melt the pollen's tough armor. The pollen grain's shell cracks open, ruptured by the swelling pair of cells within. The larger of these two cells grows, amoeba-like, out of the ruptured pollen coat and starts to burrow down between the enveloping cells of the stigma, forming a tube. Each stigma is at the tip of a stalk known as the style, and the pollen tube works its way down the style, either pushing between the cells or, if the style is hollow, flowing down the style's inner wall like a drop of oil. The second, smaller pollen cell divides and forms two sperm cells. These float down the pollen tube, carried along like rafters in a flowing river. Unlike the sperm cells of animals, mosses, and ferns, these rafters have no oars; their movement is entirely passive.

The style's length is caused by the need to hold the stigmas up where bees will bump against them. This creates a challenging odyssey for the pollen tube and, therefore, a convenient testing ground on which the plant can assess her suitors. Bees dump many pollen grains onto each stigma, so the style may have several tubes growing at once. If so, the style becomes the Kentucky Derby of plant love. The sperm cells jockey their tubes toward the ovule, which contains the plant's egg; the price of failure is the annihilation of the rider's genes. There is some evidence that vigorous plants produce fast pollen tubes, so the style's length allows the flower to select mates with a history of success. Perhaps the styles are a little longer than strictly necessary for intercepting bees, just to give the pollen stallions a good hard run.

When the pollen tube reaches the base of the style, it burrows into the fleshy ovule. Here the pollen tube releases its two sperm cells. One sperm cell joins with the egg to make an embryo; the other joins with the DNA from two other tiny plant cells to make a larger cell with a triple complement of DNA. This triply endowed cell divides, fattens up, and becomes a food storage area for the developing seed, a store that humans have put to use as wheat flour and cornmeal. Such double fertilization is unique to flowering plants; sexual union in all other creatures requires only one sperm cell and one egg cell.

The chickweed in front of my hand lens is a hermaphrodite, producing both pollen and eggs, male and female, in each blossom. Every flower contains all the necessary reproductive apparatus: pollen grains; anthers to make and store the grains; filaments to hold up the anthers, stigmas, styles; and an ovary to contain the eggs. These parts are all crowded within the cup of the flower, ringed by colored petals designed to appeal to animal eyes. Such tiny, tidily arranged complexity makes for a compelling display.

All the mandala's spring ephemeral flowers are hermaphrodites, a strategy well suited to these tiny plants that produce just a few flowers during a short, unpredictable season. By combining male and female into one flower, the plants leave open the door to self-breeding. They also spread their investment between male and female functions, increasing the chance that at least some of their genes will pass to the next generation. Other species, such as many wind-pollinated trees—oaks, walnuts, elms—use a different strategy, producing great numbers of unisexual flowers. In this case, each flower has a specialized task, either to shed pollen or to harvest pollen from the wind.

Although the mandala's plants share a hermaphroditic design, their geometry differs markedly from species to species. The *Hepatica*'s anthers grow in a thick bush around a cluster of pillarlike styles. Blue cohosh's pallid ivory flowers have globular anthers squatting around a bulbous ovary with minute stigmas. Toothwort's petals enclose a sheath around the hidden anthers. Only spring beauty has a flower somewhat like the chickweed. Its three stigmas sit atop a drooping trident, circled by five pink-tipped anthers.

This variety reflects the tastes of the plants' pollinators but is also caused by less obvious forces. Nectar robbers, for example, exert a stealthy but powerful influence over floral design. An ant has buried its head in a spring beauty flower in front of me. I use the hand lens to watch it bypass the pollen and stigma, then upend and steal the flower's sugary nectar. This robbery is the cost borne by open-cupped flowers for welcoming a diverse set of pollinators: freeloaders move in and exploit your openness. Spring beauty flowers choose the most welcoming, and therefore the most vulnerable, strategy by freely offering nectar inside an open cup that is accessible to any insect. *Hepatica* and rue anemone also produce open cups, but neither offers nectar. These nectarless flowers lose little energy to thieves, but they are also less attractive to bees. Toothwort offers nectar enclosed in a tube that excludes robbing ants but restricts the number of bees that can reach into its recesses for nectar.

The diversity of floral design is also affected by the longevity of plants and their flowers. Blooms that last for just a few days, such as those of spring beauty, are desperate for pollinators. This favors a bohemian style, risking all for the kiss of a bee. If the bee's embrace is accompanied by some ne'er-do-wells, so be it. Longer-lived flowers can be more restrained, holding back the nectar or enclosing their bloom in the knowledge that sooner or later a decent suitor will come along. The longevity of the plant that produces the bloom also factors into the economy of flowering. All the spring ephemerals are perennials that sprout from underground roots or stems. If a creeping stem lives for three decades, it can afford to be restrained in its search for pollinators. A shorter-lived root might be more willing to tolerate a few freeloaders. Both factors, the duration of the bloom and the longevity of the plant, are variants of the same theme: shorter lives must burn brighter.

Flowers therefore perform economic gymnastics as they balance losses to robbers with the need to lure pollinators. How this performance unfolds depends not just on the insects flying around the mandala but also on the ancestry of the plants. Natural selection tinkers with the raw materials provided by previous generations, so each flower's design is shaped by the particularities of its genealogy. Different plant families have different sets of equipment, constraining their acrobatics.

Hepatica and rue anemone belong to the same family, the buttercups, all of which produce nectarless flowers with open cups. Great chickweed belongs to the pink family. This family's name comes from the common name of *Dianthus*, a sweet-smelling garden flower. The flower named the color pink, and the jagged edges of its petals also gave their name to pinking shears, scissors that dressmakers use to cut zigzag edges. The "pink" plant family is named for the shears, not for the color—great chickweed inherits a tendency for serrated petals. At first glance, its ten slender white petals seem to have broken away from familial tradition. A closer look shows that the flower has just five petals, each one so deeply cleft that it seems to be double. Chickweed therefore pushed to the limit its family's preference for ornamented displays and created an illusion of extra petals.

Like all living creatures, our own lives included, the flowers layer adaptation over history, creating the tension between diversity and unity, individuality and tradition, that makes the mandala's immoderate blaze so compelling.

April 8th—Xylem

The weather has been unsettled lately, dropping sleet on one day, then blazing with hot sunshine the next. The pace of life in the mandala follows these variations. On slushy days, leaves droop and the forest is silent except for the drumming of woodpeckers. Today, the sun is out and life has quickened, with revived greenery, a dozen species of singing birds, several small swarms of flying insects, and an early tree frog chirping from a low branch.

Last week, the forest's green lay across the ground, a carpet of photosynthesis that ended at ankle height. Now the maples are unfurling leaves and dangling green flowers from branches. Like a tide rising, the green glow is reclaiming the forest from the ground up. The upward surge floods the mountainside with a sense of renewal.

Sugar maple branches hang over the mandala, and their new leaves block the sun's rays, shading the understory. Of the hundreds of spring wildflowers, only a dozen remain; the maple has snuffed their spark. But not all the trees around the mandala are in leaf. The maple's exuberance contrasts with the dour, lifeless pignut hickory that stands on the other side of the mandala. The hickory's massive gray trunk rises straight to the canopy where it holds out dark, bare branches.

The contrast between the maple and the hickory expresses an inner struggle. Growing trees must throw open the breathing pores on their leaves, allowing air to wash the wet surfaces of their cells. Carbon dioxide dissolves into the dampness before it is turned to sugar inside the plants' cells. This transformation of gas into food is the trees' source of life, but it comes at a cost. Water vapor streams out of the leaves' open breathing pores. Every minute, several pints of water are exhaled into the air by the maple above the mandala. On a hot day, the seven or eight trees whose roots penetrate the mandala send several hundred gallons of water out of their leaves as vapor. This reverse waterfall quickly dries the soil. When the supply of water is exhausted, the plant must close its breathing pores and cease growing.

All plants face this trade-off between growth and water use. But trees have another devilish layer of difficulty. By thrusting their leaves skyward they have become slaves to the physics of their plumbing systems. Inside each trunk lies the vital connection between earth and sky, soil's water and sun's fire. The rules that govern this connection are stringent.

Inside the trees' leaves, sunlight causes water to evaporate from cell surfaces and drift out of breathing pores. As vapor wafts away from wet cell walls, the surface tension of the remaining water tightens, particularly in the narrow gaps between the cells. This tension yanks more water from deep in the leaf. The pull moves to the leaves' veins, then down the water-conducting cells in the tree's trunk, finally all the way to the roots. The pull from each evaporating water molecule is minuscule, like a breath of wind tugging at a silk thread. But the combined force of millions of evaporating molecules is strong enough to haul a thick rope of water up from the ground.

The trees' system for moving water is remarkably efficient. They exert no energy, instead letting the sun's power draw water through their trunks. If humans were to design mechanical devices to lift hundreds of gallons of water from roots to canopy, the forest would be a cacophony of pumps, choked with diesel fumes or run through with electrical wires. Evolution's economy is too tight and thrifty to allow such profligacy, and so water moves through trees with silence and ease.

Yet this efficient water-lifting system has an Achilles' heel. Sometimes the rising columns of water are broken by air bubbles. These embolisms plug the flow of water. Winter weather makes these blockages more likely because air bubbles form when water freezes inside water-conducting cells. These are the same bubbles that haze ice cubes in kitchen freezers. Thus icy weather peppers the trunk with air gaps that wreck the trees' plumbing. Maple and hickory have found two different solutions to this challenge.

With its bare branches, the hickory looks wintry and inactive, but this is an illusion. Inside, the tree is building a whole new plumbing system, readying itself for the flowers and leaves that will emerge in a couple of weeks. Last year's plumbing system is useless, blocked by embolisms. So, hickory trees spend the first part of April growing new pipes. Just below the bark, a thin sheet of living cells wraps the trunk. These cells divide and create the season's new vessels. The outer layer of cells, those that lie between the bark and the sheet of dividing cells, will become phloem, a living tissue that transports sugars and other food molecules up and down the tree. The new cells formed on the inner side will die and leave their cell walls to become the xylem, or wood, that conducts water up the trunk.

Hickory xylem tubes are long and wide. These pipelines offer little resistance, so the flow of water is prodigious when the tree finally leafs out. But the width of these tubes makes them particularly vulnerable to blockages by embolisms. Once blocked, they become useless and, because the tree has relatively few of these wide conduits, the flow of water drops significantly with just a few embolisms. This design means that hickories must delay the growth of their leaves until the danger of frost is past. The trees miss out on the warm sunny days of spring, but they recoup these losses when their pipelines are thrown open later in the season. The hickory is therefore like a sports car—kept off the road by ice until late in the spring, but outstripping all rivals on warm summer days.

Hickory trunks have one more problem. Their wide, long xylem tubes are weak, like thin-walled straws. These tubes cannot hold up heavy branches or cope with the force of wind pulling on leaves. Therefore, later in the year, after the springtime xylem has grown, the tree grows thick-walled, smaller-bored xylem vessels. This summer-grown xylem provides the structural support that the water-carrying vessels lack. The yearly alternation is visible in cut hickory wood as a "ring porous" pattern of wide porous cells separated by denser wood.

If hickories are sports cars, then maples are all-wheel-drive passenger cars. Their xylem is frost-resistant and lets them leaf out weeks before the hickories. But, come summertime, maples will lag behind hickories in their ability to carry water and thereby feed on sunlight. The maples' xylem cells are more numerous, shorter and narrower than the hickories', and they are separated by comblike plates. Unlike the broad, open tubes of hickory, the maple's design confines embolisms to the small cells in which they form. Because maples have so many small tubes, each embolism blocks just a tiny fraction of the trunk. Unlike the ringed patterns in hickory wood, maple wood is more uniform, showing a "diffuse porous" pattern. These differences are visible in furniture and other wood products—maple is smooth-grained, whereas hickory has regular rows of pinholes.

The maple has one more physiological trick to help it cope with embolisms. Sugary sap rises forcibly up maple trunks in the early spring, flushing out air and restoring the old xylem's integrity after winter's hard freezes. Maples can therefore use old xylem for extra water-carrying capacity, whereas hickories are restricted to the current year's growth. The maples' springtime flow of sap is powered by cycles of nighttime freezes and daytime thaws in their twigs. This explains why sap flows heavily in some years and hardly at all in others. When temperatures see-saw between sharp nighttime frosts and sun-warmed days, sap flows prodigiously; when the weather is uniformly tepid, the flow is stanched.

The contrast between the leafy maple and the somber hickory comes down to a matter of plumbing. The trees seemed at first to be prisoners of the unyielding laws of physics. The constraints imposed by water's evaporation, flow, and freezing circumscribe their lives. But trees are also masterful exploiters of these same laws. Evaporation is the cost that trees pay for opening their leaves, but evaporation is the force that silently and effortlessly moves hundreds of gallons of water up tree trunks. Likewise, ice is the enemy of springtime xylem, but ice powers the maples' early flow of sap, again without cost to the tree. In two different ways, both maple and hickory have turned the tables on their constraints and turned adversity into triumph.

April 14th—Moth

A moth shuffles its tawny feet over my skin, tasting me with thousands of chemical detectors. Six tongues! Every step is a burst of sensation. Walking across a hand or a leaf must be like swimming in wine, mouth open. My vintage meets the moth's approval, so his proboscis unfurls, rolling down from between the bright green eyes. Unrolled, the proboscis juts straight down from the moth's head, like an arrow pointing at my skin. At the point of contact, the proboscis's rigidity softens and the tip flops backward, pointing between the moth's legs. I feel cool wetness as the moth slaps the tip around, seeming to search something out. I lean toward my finger, squinting through a hand lens in time to see the tip worm itself into the groove between two ridges of my fingerprint. The proboscis stays lodged in this furrow, and fluid flashes back and forth in the pale tube. The sensation of moistness continues.

I watch the moth feed for half an hour and discover that I cannot dislodge my guest. At first I hold my finger still, cautiously moving only my head. After several minutes my body protests at such stiffness, so I move the finger. No reaction. I wave the finger, then blow on the moth. Again, the moth continues its work. Pokes with my pencil end fail to stir the animal. A large fly also visits and dabs my hand with wet kisses from its toilet-plunger mouth. This bristly fly shows more normal insect reactions and takes flight as I lean close. The moth, however, sticks like a tick.

The moth's antennae hint at the cause of the vigorous attachment to my finger. The antennae arch out of the head, reaching forward by nearly the length of the moth's body. Closely spaced ribs jut out from each antenna's spine. The moth is thus crowned with two threadbare feathers. These plumes are covered with velvety hairs. Each hair is peppered with holes that lead into a watery core in which sits a nerve ending, waiting for the right molecule to bind to its surface and trigger a response. Only males have such exaggerated antennae. They comb the air for scent released by females and fly upwind, guided to a mate by their enormous feathery noses. But finding a mate is not enough. The male must provide a nuptial offering to his mate. My finger provides him with an essential ingredient for this gift.

Diamonds may be the crystal of choice for wooing humans, but moths seek a different, altogether more practical mineral, salt. When the moth mates he will pass to his partner a package containing a ball of sperm and a packet of food. This food is generously seasoned with sodium, a precious gift that looks forward to the needs of the next generation. The female moth passes the salt to the eggs and thus to the caterpillars. Foliage is deficient in sodium, so the leaf-munching caterpillars need their parents' salty bequest. The moth's arduous attachment to my finger prepares him for mating and will help his offspring to survive. The salt in my sweat will make up for deficiencies in caterpillar diets.

The morning is sunny and comfortably warm. Summer's heat has yet to arrive, so I am barely sweating. This makes the moth's task harder and provides a poor chemical mix for his gift. Copious sweat would be much better. Human sweat is made from blood with all the large molecules removed, like soup passed through a sieve. The blood fluid passes out of our vessels, seeping around the spaces between our cells and into the coiled tubes at the bottom of sweat ducts. As the fluid passes up the sweat ducts the body pumps sodium back into its cells, reclaiming the valuable mineral. The faster the sweat moves, the less time the body has to recapture the sodium, so when we pour with sweat there is little difference between the mineral mixture in our sweat and that in our blood. We are literally sweating blood, minus the lumps. When the sweat moves out of us sluggishly, we produce fluid that has less sodium and proportionally more potassium, a mineral that the body expends little effort on reabsorbing. Plant leaves have plenty of potassium, so male moths are not interested in it and void any that they suck up with the sodium. Some of what the moth is taking from my skin will therefore pass into his feces and thence back to the soil.

Despite providing the moth with barely a trickle of the wrong flavor of sweat, I am a mammal worth clinging to. Humans are one of the few animals to use sweat as a cooling mechanism, so salty skin is seldom found in the mandala. Naked salty skin is rarer yet. Bears sweat and so do horses, but their bounty is hidden under a layer of hair. Horses never visit the mandala. Bears are very rare, although remains in local caves show that they were common before the arrival of gunpowder. Most other mammals sweat only on their paw pads or lip margins. Rodents don't sweat at all, perhaps because their small bodies make them particularly vulnerable to dehydration.

Blood fluid oozing from pores is therefore an unusual treat at the mandala. My skin's meager sweat is a feast compared to the scarcity of sodium in the forest. Rain puddles sometimes are worth sucking at, but they are seldom rich in sodium. Feces and urine are saltier, but they dry quickly. I am the best bet today. Not wanting to carry the moth out of the forest when my sit at the mandala is over, I must pry his grasping feet from my skin, then run away.

April 16th—Sunrise Birds

A peach stain soaks into the darkness on the eastern horizon, then the whole dome of the sky lightens, bleeding from darkness to pale luminosity. Two repeated notes ring through the air; the first is clear and high, the second is lower and emphatic. These tufted titmice keep up their rapid two-part rhythm as a Carolina chickadee starts a whistled melody, four notes that fall and rise like a nodding head. The peach spreads up from the horizon, and a phoebe calls with a whiskey-and-cigarette-roughened voice, rasping out its name, *phwe-beer*, like a broken bluesman.

As the sky's pallor brightens, a worm-eating warbler rattles a caffeinated castanet. The dry buzz unleashes a confusion of songs from all directions, a jumble of tempos and timbres. The black-and-white warbler wheezes lazily, *whee-ta whee-ta*, from an upside-down perch under a tree limb. The hooded warbler rings out from a sapling, twirling the notes twice around to gather speed, then flinging them to the sky, *wee-a wee-a WHEE-TEE-O*. From the west comes a yet louder song. Three rich tones wash over the forest like repeated waves, then break down into rippling eddies. The tin-whistle song of the Louisiana waterthrush seems inspired by the flow of the streams along which it lives, yet the song's cadence and volume carry the sound above the water's roar.

Peach turns to pink, and color spreads wider over the horizon. The sky's vault brightens enough to reveal the partly closed chickweed flowers in the mandala and to give form to the boulders and stones that define the mandala's edge. As the world fades into view, Carolina wrens sing, vying with the waterthrush to produce the loudest song in the woods. The wrens sing year-round, but today I hear them with fresh ears, the familiarity of their song stripped away by the springtime flush of sound. No other bird, save the now-departed winter wren, can match the vigor of their acoustic attack or the exuberance of the uncoiling energy in their song.

The wren's music is answered by a Kentucky warbler from farther down the slope. The warbler echoes the wren's theme and tone but holds back, like a diver bouncing endlessly on a board, never daring to take the plunge. Then another song bursts from the canopy, lisping like a black-and-white warbler, but the song breaks out of the pattern, accelerates, then twitters. I cannot identify the bird and, more frustratingly, cannot find it with my binoculars. Perhaps this is the dawn "flight song" of a warbler? These flight songs are out-of-character virtuoso solos given during arching flights high above the forest. They are seldom recorded and, in my limited experience, are highly variable. What role they play in the birds' lives is unknown but, if nothing else, they must provide a rush of creative release for birds that spend the rest of their day repeating just a few syllables.

Woodpeckers add their boisterous voices to the performance. First the red-bellied woodpecker lobs its quivering cry across the mandala, then an answer flies back, the maniacal laugh of the pileated woodpecker. Blue jays punctuate the woodpeckers' volley with alternating rasps and whistles. As the glow in the sky intensifies, half a dozen goldfinches fly eastward, bouncing in the air just above the forest canopy, like flung stones skimming over water. Each bounce is accompanied by a twitter, *ti-ti-ti, ti-ti-ti.*

The whole sky flashes pink for a moment, then yellow surges up from the east, brightening the mandala. Color sinks back to the horizon again, leaving milky light across the rest of the sky. A red-eyed vireo greets the glow with his regularly spaced bursts of whistling. Some bursts end on a rising note, "where am I?"; others conclude on a low note, "there you are . . ." The vireo questions the forest, then answers over and over, lecturing into the midday heat when other birds have retired from the podium. As befits his professorial temperament, the vireo seldom descends from the heights of the canopy and is usually detected only through his bright, repetitious song. The vireo is joined by a brown-headed cowbird. Cowbirds are brood parasites, laying their eggs in the nests of other birds. This emancipation from parental duties leaves the cowbirds free to pursue the pleasures of courtship. The male's song has taken him two or three years to perfect and sounds like molten gold falling, solidifying, then ringing out as it strikes stone. A burst of precious liquidity combined with the ringing of metal.

The heavens shine blue now, and the sunrise's colors have faded to a pastel belt of cloud in the east. A northern cardinal chips loudly on the slope below the mandala, each note like struck flint. These crisp calls are the counterpoint to turkey gobbles rising up from the valley below. The forest has muffled the turkey's distant sounds, adding what Thoreau called the "voice of the wood nymph" as the sound is bounced and squeezed through the vegetation. We are in turkey-hunting season, so the gobbles are as likely to be gobble-mimicking humans on a gastronomic quest as they are real turkeys searching for love.

The fading dawn colors revive momentarily, and the sky shines with lilac and daffodil, layering colors in clouds like quilts stacked on a bed. More birds chime into the morning air: a nuthatch's nasal *onk* joins the crow's croak and a black-throated green warbler's murmur from the branches above the mandala. As the colors finally fade under the fierce gaze of their mother, the sun, a wood thrush caps the dawn chorus with his astounding song. The song seems to pierce through from another world, carrying with it clarity and ease, purifying me for a few moments with its grace. Then the song is gone, the veil closes, and I am left with embers of memory.

. . .

The thrush's song flows from the syrinx buried deep in his chest. Here membranes vibrate and squeeze the air that rushes out of the lungs. These membranes circle the confluence of the bronchi, turning a toneless exhalation into sweet music that ascends the trachea and flows out of the mouth. Only birds make sound this way, using a biological hybrid between the flute's swirling tube of air and the oboe's vibrating membranes. Birds change the texture and tone of their songs by adjusting tension in the muscles that wrap the syrinx; the thrush's song is sculpted by at least ten muscles in the syrinx, each one shorter than a grain of rice.

Unlike our voice box, the syrinx offers little resistance to the flow of air. This gives small birds the ability to ring out louder songs than the huskiest human. But despite the efficiency of the syrinx, birdsong seldom carries farther than a stone's throw. Even the turkey's explosive gobble is quickly swallowed up by the forest. The energy that propels the sound is easily absorbed and dissipated by trees, leaves, and the sponginess of air molecules. High-pitched sounds are more easily absorbed than bass notes, whose long wavelengths let them flow around obstacles rather than bounce away. The beauty of birdsong, especially descant birdsong, is therefore a blessing available only at close range.

Not so the sun's gift. The photons that created this dawn have traveled one hundred and fifty million kilometers from the surface of the sun. But even light can be slowed and filtered. This slowing is most dramatic inside the sun's belly, where photons are born from the fiery union of pressurized atoms. The sun's core is so dense that it takes ten million years for a photon to struggle to the surface. Along the way, the photon is continually blocked by protons, which absorb the photon's energy, hold it for a moment, then release the energy as another photon. Once the photon finally bursts free from millions of years trapped in the sun's molasses, it zips to earth in eight minutes.

As soon as photons reach our atmosphere their paths are again strewn with molecules, albeit molecules that are millions of times less densely packed than those in the pressed mass of the sun. Photons come in many colors, and some colors are more vulnerable to being impeded by the atmosphere. Red photons have wavelengths that are much longer than the size of most air molecules, so, like a turkey gobble in a forest, they flow easily through the air and are seldom absorbed. Blue photons have wavelengths that more closely match the size of air molecules, so this short wavelength is absorbed by the air. An air molecule that absorbs a photon jiggles with the excitement of the ingested energy, then pops out a new photon. The ejected photon is shot out in a new direction, so the tidy stream of blue photons is scattered into a ricochet of light. Red light is not absorbed and scattered, so it flows straight on through. This is why the sky is blue; we are seeing the redirected energy of blue photons, the glow of billions of excited air molecules.

When the sun is overhead, photons of all colors reach our eyes, even though some blue ones are redirected along the way. When the sun is low on the horizon, photons have a sloping path to cut through the air, so more blue light is stripped out. The red dawn light bathing this Tennessee mandala was therefore born in the blue morning skies over the Carolina mountains to the east.

The light and sound energies washing over the mandala find a point of convergence in my consciousness, where their beauty quickens a flame of appreciation. There is convergence also at the start of the energy's journey, in the unimaginably hot, pressurized core of the sun. The sun is origin of both the dawn's light and birds' morning songs. The glow on the horizon is light filtered through our atmosphere; the music in the air is the sun's energy filtered through the plants and animals that powered the singing birds. The enchantment of an April sunrise is a web of flowing energy. The web is anchored at one end by matter turned to energy in the sun and at the other end by energy turned to beauty in our consciousness.

April 22nd—Walking Seeds

The springtime flush of flowers is over. A few chickweed flowers and a geranium are all that is left of the month's glory. Spent flowers rain down from above, bringing to earth evidence of the maple and hickory trees' prodigious reproductive efforts. Hundreds of maple and hickory flowers are lying within the mandala. Unlike the gaudy blooms of the spring ephemerals, these tree flowers are bland and unassuming, with no obvious petals or colored adornments. This extreme Puritanism of dress suggests that sex among the mandala's trees is a business very different from the ephemerals' effusive festival of nectar and color. These trees have no one to impress. Wind carries their pollen, so insect eyes and palates need not be bribed; flowers can be stripped down to their utilitarian essentials.

Pollination by wind is a particularly useful strategy for early-blooming trees. The spring ephemerals live in a relatively warm, sheltered microclimate, yet they struggle to find pollinators. The tree canopy's microclimate is more exposed and is even less friendly to the insects of early spring. Wind is not in short supply, however. Maples and hickories have therefore broken the ancient contract with insects, using physical rather than biological methods to transport their pollen. Increased reliability comes with an unfortunate decrease in precision. Bees deliver pollen directly to the stigma of the next flower. Wind does not deliver anything. Rather, it disperses whatever is caught in its motion, much to the distress of both flowers and human sinuses.

Wind-pollinated plants must therefore release vast drifts of pollen. They are like castaways stranded on an island, throwing millions of bottles into the water for want of a dependable postal service.

Unlike the hermaphroditic wildflowers, maple and hickory produce two kinds of flower, male and female. Male flowers dangle from twigs so that the slightest movement of air will stir them. Maple trees hang clusters of these flowers from wiry filaments. Each filament is a centimeter or two long and ends in a tuft of anthers, pollen-producing structures that look like tiny yellow balls about the size of a comma on this page. Hickory's anthers are strung on fuzzy garlands called catkins, each one about as long as a finger. In both species, the anthers nestle in groups under small umbrellas, presumably to stop the rain from washing pollen away. The female flowers are more stubby, having no need to cast large quantities of pollen into the wind. Their stigmas intercept wind-borne pollen to start the fertilization process. Little is known about the aerodynamics of the stigmas, but they seem to be placed in the windiest parts of the plant and to be designed to encourage air to curl around them, forming eddies that slow the air and thus deposit pollen grains.

By this time in the season, male flowers have shed their pollen and, their task completed, have been discarded by the trees, leaving the mandala covered with tangles of yellow-green filaments and catkins. But the work of the female flowers has just begun. The fertilized eggs inside these flowers will take months to mature into fruit. Mature hickory nuts and maple seeds will not be ready to drop until autumn.

Wildflowers do not have the luxury of months of summer sun to fuel their fruiting. Most spring ephemerals fruit just a few weeks after their bloom, completing all their reproduction for the year before summer's thick tree canopy chokes off the light. I walk around the mandala's edge to search for the *Hepatica* plant whose flower I watched open in March. I find it just behind the spicebush, its liver-leaves splayed wide and the flower stalk holding up a bouquet of fat green torpedoes, each one the size of a small pea. Several of these fruits have tumbled to the ground, revealing a blunt white nipple at their base, a bulbous center, and a sharply tapered tip. This sharp tip is all that remains of the style, the short stalk that supported the stigma. The green swelling is the ovary wall, which now encloses a fertile seed.

An ant approaches one of the fruits, palpates it with her antennae, and crawls on top of the fruit. She scurries back onto the leaf litter, grasps the fruit, then abandons it. Another ant repeats the process a few minutes later. Each time the fruit moves a few millimeters, but the ants then leave. Half an hour passes and more ants walk by, ignoring the fruit. Then a large ant appears, tickles the fruit with her antennae, and seizes it with the hooked mandibles that jut from either side of her mouth. The fruit is as large as the ant, but she raises it high above her head, mouthparts firmly planted in the fruit's blunt white end. She sets off for the center of the mandala, stumbling over maple flower stems, recovering, falling into leaf crevices, crawling onward. Her path is tortuous, circling back to bypass gashes in the leaf litter, walking backward through tangles of catkins. I get caught up in her struggle and exhale sharply with relief when she reaches a penny-sized hole in the litter and ducks down. I peer into the ant hole and see the green glow of the fruit being jostled and rotated by a small group of ants. Gradually the glow fades as the fruit is swallowed by the ground, a foot away from where it fell.

The *Hepatica* fruit's odyssey is part of a larger saga, linking the stories of forest ants to those of spring ephemerals. The white nipple at the tip of the *Hepatica* fruit is an elaiosome, a fatty treat cooked up by the plant especially for ants. Such rich food is seldom found in convenient, undefended packages, so ants quickly carry elaiosome-bearing fruits to their nests, where the food parcel is chopped up and fed to the colony's larvae. The next generation of ants will be partly made from *Hepatica* flesh. Once the elaiosome has been removed, the ants dump the inedible seed into their nest's compost heap. Thus the fastidious tidiness of the ant colony places the seed in loose, fertile compost, the perfect place for germination.

Not only do ants sow seeds in convenient places, they also help move them away from their parent plants and into potentially unoccupied spaces. Most ants move the seeds of spring ephemerals a few feet, rarely more than a stone's throw from the parent. This is enough to avoid competition with mom, but such short dispersal distances are hard to reconcile with what we know about the history of the ephemerals. Many ephemerals have populations that cover the entire extent of the temperate forest in eastern North America, starting in Alabama and stretching all the way into Canada. Yet sixteen thousand years ago this temperate forest was squeezed into a few pockets on the Gulf of Mexico. The last ice age covered the rest of the east with ice or, in the more southern areas, with the kind of boreal forest now found only far north in Canada. Spring ephemerals have therefore moved from Florida to Canada in sixteen thousand years. But if post-ice-age ants behaved in the same way as modern ants, the ephemerals would have moved just ten or twenty kilometers since the retreat of the ice, not the two thousand kilometers they evidently have managed. Either today's ants are shadows of the great sprinting ants of yesteryear, which is unlikely; or the fossil and geological evidence for the ice ages is a mirage, which is even less likely; or our understanding of seed dispersal is incomplete, and spring ephemerals have some unknown method of long-distance transport.

Until recently candidates for this "mystery disperser" all seemed rather weak. Freak windstorms carrying *Hepatica* seeds to Canada? Unlikely. Mud under the toenails of migrant birds, or seeds carried in bird bellies? Possible, but most migrants have passed through southern forests before the ephemerals have set seed. Toadshade trillium plants set seed so late that migrant birds have started their return journey and would carry seeds the wrong way. Rodents or other herbivores carrying seeds in their guts? Dismissed out of hand: they grind up seeds in their mouths, then destroy them during digestion.

Ecologists have dubbed the mismatch between the ephemerals' rapid advance and their seemingly poor dispersal abilities "Reid's paradox," after a nineteenth-century botanist who encountered a similar problem with the spread of oaks across postglacial Britain. Philosophers and theologians love paradoxes, regarding them as honorable signposts to important truths. Scientists take a dimmer view, having learned from experience that "paradox" is a polite way of saying that we are missing something obvious. The resolution of the paradox will likely show one of our "self-evident" assumptions to be embarrassingly false. Perhaps this is not so far removed from a philosophical paradox. The difference lies in the depth of the false assumption: relatively shallow and easily uprooted in science, deep and hard to dislodge in philosophy.

The false assumption underlying Reid's paradox may not be buried at all but perhaps lies on the leaf litter in mandalas all across the continent. Deer droppings, which, like the feces of rodents, we had assumed contained no viable spring ephemeral seeds may turn out to be the paradox's solution. This resolution meets the criterion for classic scientific paradox-solving: a simple experiment with a "why didn't anyone think of that before?" answer. Step one, collect deer droppings from the woods; step two, search the droppings for seeds; step three, plant the seeds, watch them grow, and conclude that "ant-dispersed" seeds are misnamed. Perhaps "ant-jiggled" and "deer-lobbed" would be a better description, because deer, it turns out, can transport seed many kilometers. Ants manage mere centimeters. And what of the other herbivorous mammals that we also dismissed as potential carriers of seed? No one has stooped to pick up after them to find the answer. We have a lot of dung sifting ahead of us.

Whatever the sifters find, we can already conclude that we were premature to categorize many spring ephemerals as "ant-dispersed," even giving their relationship a heavyweight label, myrmecochory. The reality of seed dispersal is more complex and seems to depend on scale. At a small scale ants are indeed the main dispersers. They excel at collecting the seeds and planting them in prime locations. Deer are much less careful gardeners. Therefore, from the perspective of an individual seed, there is no better fate than to be discovered by an ant. However, at a larger scale mammals are vastly more important than ants. The occasional successful long-distance transport of a seed by deer can found a new population and vault the species into a previously unoccupied forest. From the perspective of the whole species, footloose deer are more important than fastidious, shuffling ants. Without deer, the ephemerals would be confined to a small strip of forest on the Gulf Coast. Instead, they have hitchhiked across the continent.

The newfound importance of deer calls into question the function of the elaiosome. We have assumed that the oily appendage was designed by natural selection to attract ants, thereby helping to place the seed in convenient potting soil. This explanation is still likely true, in part. Ants are, after all, the best sowers of seed, and natural selection will exalt any characteristic that helps pass on genes to the next generation. But selection also favors those characteristics that will carry genes to the four winds. Evolution commands not just "multiply" but "go forth and multiply." Any mother that does not launch some of her children on long voyages will lose out in the long run. This is particularly true in species whose history has been marked by the recolonization of vast areas of habitat. Nearly every *Hepatica* flower in North America is the descendant of a successful long-distance disperser. We should expect to find itchy-feet genes, characteristics that make it more likely that a seed will be deposited far from its parents. The elaiosome, therefore, may be partly designed for this purpose, titillating supple deer lips, encouraging them to pluck the hopeful fruit.

The paradoxical lives of the ephemerals have had new layers of complexity folded into them by the arrival of Europeans. We have chopped the forest into pieces, making it harder for ants to move seeds around. At the same time, deer populations have crashed, then boomed. The ant-deer balance had tipped. How will the ephemerals respond? Or, can they respond? Large numbers of deer can turn the blessing of dispersal into the curse of overbrowsing. Sustained heavy munching will wipe out ephemerals, making moot our speculations about their response to changing natural selection.

Now, a third weight is added to the balance. Imported fire ants have invaded southern woodlands and are moving north. They thrive in disturbed areas, making them particularly common in forests that are already reeling from the effects of fragmentation. Fire ants gather elaiosome-bearing fruits but are poor dispersers, depositing seeds next to the mother plant, thus dooming seedlings to a childhood spent competing with a larger relative. These competitions usually end with the seedling's death. Fire ants may also be predators, eating the whole fruit instead of just the elaiosome. This invasion of foreign ants has the potential to undermine the relationship between the elaiosome and all its native dispersers, making the oily gift a liability instead of the asset it has been for millennia. Ephemerals may be caught in a race between natural selection and extinction. Either they will adapt to new conditions, or their numbers will dwindle in the face of a new reality for which they were unprepared.

The spring ephemerals' passage through the tumult of the ice age shows that they can readily adapt to changing ecological winds. But the ice age was a storm that came and went over thousands of years. Now the plants are faced with unpredictable changes that squall over them in just decades. The ecologist's paradox has become the conservationist's prayer. This mandala may be part of the answer to that prayer, a relatively unfragmented, uninvaded piece of forest where the old ecological rulebook has yet to be entirely torn up and blown away. These ants, these flowers, these trees contain the genetic history and diversity from which the future will be written. The more wind-tattered pages we can hold on to, the more materials evolution's scribe will have to draw upon as it reworks the saga.

April 29th—Earthquake

The earth's belly rumbles mightily. Intestines of stone shudder past one another, untwisting their tension, grinding into relaxation. The distress is centered sixty miles away, twelve miles below the skin. As the pent-up energy of stressed rock is released, some of the fury spreads outward in waves of shifting earth.

The compression waves arrive first, roaring. Like a herd of diesel trains, the sound rips over the land, jarring us into confusion from predawn sleep. The sound spills out of the earth, washes over us for a few seconds, then flies on. These compression waves rip through the earth at more than a kilometer every second. A pause follows, then surface waves hit, bucking and shaking the house. The waves combine horizontal and vertical motion, squeezing and shearing simultaneously. Like small boats tacking against ocean swell, houses are twisted and rolled in the geological storm. In a big swell, houses break apart, unable to withstand the wrenching stresses.

We are fortunate. The swell is modest and our house stays upright. The roar outside the house is replaced by chiming and clanging from within. Framed pictures hanging from the wall act as pendulums; when the earth lurches the house to one side, the heavy frames remain stationary, stilled by inertia. Then the wall returns, bang!, jumps away, snaps back, bang! Keys jingle, glasses knock together in a toast to the quake, plates slide and clang. Everything tied to the earth is moving, all else is still or slowed, but our eyes deceive us and tell us that the contents of the house are dancing within stationary walls. The shakes continue for about fifteen seconds, then die out in trembling fits.

The inertia of suspended objects is put to use in measuring the strength of earthquakes. A pen hung from the weighted bob of a pendulum will scribe the earth's movements on graph paper held below its point. When the quake strikes, the pen remains still, but the paper and the pendulum's frame both move, causing the pen to record the extent of the movement. Some seismographs have pendulums three stories tall, recording every little quiver in the ground below.

The calibrated scratchings of dangling pens give us the Richter scale. This morning's earthquake scored 4.9 on the scale, about the same magnitude as a small nuclear weapon or one thousand times the strength of a powerful quarry blast. Because the Richter scale is logarithmic, the actual amount of energy contained in an earthquake increases exponentially with the numbers on the scale. An earthquake scoring 3 on the scale is minor, a 6 will do some damage, 9 is devastating, and a Richter 12 quake is so powerful that it would crack the earth in two, or so we are told.

I hurry to the mandala at first light, anxious to witness the geological consequences of the earthquake. Mountains are dynamic beings, so I hope to see some rolled rocks or cracked cliffs. But things are just as I left them. The mandala appears entirely unmoved. If change has occurred, it is beyond my perception. The sandstone boulders sit like old monks, beyond stillness in the depth of their contemplation.

I have run up against a break, a discontinuity in the nature of reality. The biological drama that plays around and over the mandala's stones keeps time in seconds, months, or centuries and measures physical scale in grams or tons. Geological reality ticks in millions of years and weighs in billions of tons. It seems that I am extraordinarily unlikely to see geology in action at the mandala, even after an earthquake. The tempo and scale of geology are incommensurable with biological experience.

We bury our incomprehension in the usual way, with words. The mandala's rocks are about three hundred million years old, built from the huge river of sand that flowed from a still older mountain chain to the east. The earth's crust has, grain by grain, disassembled and rebuilt itself over and over to the rhythm of millions of millennia. These are preternatural ideas, beyond the nature of our experience or imagination.

The earth's slow movements seem to exist in another realm, separated from life by a wide chasm of time and physical scale. This is challenge enough for our minds. But the most unfathomable truth about the chasm is that there is a thread across, a thin connection from life's moment-by-moment to the impossible longevity of stone. The thread is woven by life's persistent fecundity. Tiny strands of heredity join mother to child and combine to stretch back billions of years. The strands spool year by year, sometimes branching into new lines, sometimes ending forever. So far, diversification within the thread has kept pace with extinction, and the mortal biological fleas on the immortal stony gods have bought a contingent immortality of their own. But every strand in the rope is a race between procreation and death. Life's generative force has been strong enough to win this race year by year for millennia, but final victory is never guaranteed.

The mandala sits at just one point along this thread. The rest of the chasm is bridged by the ancestors and descendants of the species here. None of these living creatures will truly experience the vastness of geological time. It is therefore easy to forget or ignore this vastness, assuming that our physical setting is fixed, "set in stone." I sit at the mandala under a bluff that is now the western edge of the Cumberland Plateau. The land here is made from sandstone, with limestone farther down the slope. Water from this mountainside runs into the Elk River, then on to the Gulf of Mexico. These realities form the seemingly firm walls of the mandala's world. But walls turn out to be veils. Behind the veils, across the chasm, the world is in motion. The mandala sits on an old river delta that, in turn, sits on an ancient sea floor. All this was uplifted and eroded. Oceans, rivers, and mountains changed places in a dance of terrifying magnitude. The mandala was shaken by an infinitesimally small finger twitch of the dance last night, a reminder of the overwhelming otherness of the physical earth.

May 7th—Wind

A gumball-sized *Mesodon* snail slides its gray body across the leaf litter, then climbs a twig. Halfway up the twig it lurches sideways and falls to the ground, felled by the wetness that slicks every surface in the mandala. Two days of storms have pushed water into every crack and pore. Saplings are weighed down by the burden of droplets, and the remaining ephemeral wildflowers are squashed down, defeated by the incessant pounding of rain. A patch of mayapple just to the west of the mandala has been leveled, as if crushed by a huge roller. Although it is well past dawn, the sky is dark and casts a dim light that deepens the wetness. The damp air oozes around the mandala, merging sky and forest. The leaf litter seems to have no upper surface; the rotting leaves simply bleed upward and turn into dark wet air.

The storms came with strong winds, some of which swirled into tornadoes. None of the columns of vengeful air touched the mandala, but the forest floor is strewn with the evidence of agitation in the canopy. Freshly torn leaves festoon the litter. Cracked twigs and fallen branches are tangled in the understory. The wind's force has yet to die down. It surges across the forest in pulses, setting the trees into violent motion. The canopy protests with a loud hiss, the sound of millions of pummeled leaves. The forest groans and cracks as tired wood fibers are pushed beyond their endurance.

The air is quieter at ground level. Strong breezes rush past me, but it is calm enough for mosquitoes to circle my arms and head, weaving in and away as they plot their attack. The mosquitoes and I sit in the middle of a dramatic gradient of physical energy. The surface of the canopy is the shore against which the air beats itself, crashing wave after wave onto the treetops. The shrub layer of the forest, where I sit, is heavily buffered by the trees above and receives only feeble eddies from the breakers pounding the canopy. The mandala's surface is calmer yet. Snails feel hardly a breeze as they graze on the leaf litter. No insect or snail is active in the canopy today; only a few brave the gusts below, but life continues as usual in the leaf litter.

Trees are ill suited to absorb the force of the wind. Leaves are designed to intercept as much sun as possible. Unfortunately this also makes them excellent wind catchers. The saillike surfaces of leaves are pulled leeward by the flow of air. Leaves and twigs do not stretch much, so the pull is transmitted to the rest of the tree. As the wind gets stronger, leaves start to flutter. A fluttering leaf creates more drag than a stiff leaf, so the pull on the tree increases sharply. The force of tens of thousands of leaves dragging in the wind is accentuated by the height of the tree's crown. The trunk acts as a lever, turning the tree into a huge crowbar. The wind pulls at one end, the trunk multiplies the force, and snap! the tree is shattered or uprooted.

Natural selection does not allow trees to take the obvious way out, namely to abandon the lever arm and hug the ground. Competition for light among plants in the forest forestalls this possibility. Any tree that fails to grow a tall trunk will be unable to gather much sunlight and will leave few, if any, offspring. Trees therefore grow as tall as their supporting architecture will allow, each individual tree reaching up to secure an unshaded spot in the canopy. A second solution to the problem of wind would be to stiffen the trunk, toughen up the twigs, and turn the leaves into solid plates. This is the human approach: our solar panels and satellite dishes are firmly anchored and flap in the wind only when things go wrong. But this approach is costly. Solid trunks and leaves would require a hefty investment in wood. Platelike leaves would also be less effective at photosynthesizing, having lost their gauzy openness to light and air. Such leaves would also take longer to make, delaying the tree's springtime growth. Bulking up is therefore a poor solution.

The tree's answer to the wind's force echoes the Taoism of the lichens: don't fight back, don't resist; bend and roll, let your adversary exhaust herself against your yielding. The analogy is reversed, for the Taoists drew their inspiration from nature, so "the Tao is Tree-ist" is more accurate.

In moderate winds, leaves bend back and flutter. As the wind's force increases, leaves change their behavior and absorb a portion of the wind's strength, using it to furl into a defensive posture. The leaves fold onto themselves, rolling their margins to the center. They take on shapes of strange fish, shedding air from their aerodynamic surfaces. The compound leaves of hickories fold each leaflet to the central stalk, forming a loosely rolled cigar. Air rushes past, its death grip loosened. As the wind abates, the leaves spring back, unrolling into sails again. Lao Tzu reminds us: "Grass and trees are pliant and fragile when living, but dried and shriveled when dead. Thus the hard and strong are the comrades of death; the supple and the weak are the comrades of life. A weapon when strong is destroyed; a tree when strong is felled."

The trunk also yields to the wind's push rather than resisting like a rock. It is designed to stretch and flex, absorbing energy in the microscopic cellulose fibers from which wood is woven. The fibers are arranged in coils, so each one acts as a spring. The coils are layered over one another, forming the water-carrying tubes that run up and down the trunk. Each tube has many coils, and each coil is wound at a slightly different angle. The result is a trunk filled with springs, each spring designed to exert its maximal pull at a different degree of stretch. The tightly wound springs resist strongly as the wood is first stretched. Looser springs take over as the tension increases and the tight springs fail.

I look out across the forest and see nothing but trunks in motion. They scissor past one another, bending alarmingly as their crowns surge back and forth. Despite their elegant accommodation and avoidance of the wind's power, there is a good chance that some of them will fall. Within five paces of the mandala there are two large fallen trees. Judging from their freshness, they likely fell within the last year or two. One, a hickory to the east, was uprooted. The other, a maple to the north, snapped its trunk four feet from the ground. Both trees were smaller than those that surrounded them. Perhaps their vigor was sapped by shading from larger competitors? If so, they would have grown little new wood, and fungi may have invaded the weakened trunks and roots, chewing up the cellulose coils. Bad luck may also have been involved. Either tree could have been hit by a particularly strong gust, and the hickory was growing amid root-blocking boulders. Whatever the particularities of their history, these fallen trees have now started the next part of their journey through the ecology of this old-growth forest. Fungi, salamanders, and thousands of species of invertebrates will thrive in and under the rotting trunks. At least half a tree's contribution to the fabric of life comes after its death, so one measure of the vitality of a forest ecosystem is the density of tree carcasses. You're in a great forest if you cannot pick out a straight-line path through fallen limbs and trunks. A bare forest floor is the sign of ill health.

Today the forest floor is strewn not only with fallen trees and limbs but with green maple helicopters, the discarded immature fruits whose seeds were defective or whose stems were too weak. The seed embedded in each fruit was fertilized by sperm from a wind-borne pollen grain. The fruit is an airfoil, so as it spins it creates an upward push and its descent is slowed, increasing the distance it can travel. The wind is therefore the maple's goddess of both sexual union and childhood wanderlust.

The diversity of shapes of the maple helicopters scattered on the mandala suggests that maples are not passive recipients of the wind goddess's whims; trees have the potential to mold themselves to her character through natural selection. Variation in fruit design may lead to evolutionary adaptation: those helicopter shapes best suited to the nature of wind in their corner of the world will survive and prosper. Even without such evolutionary change, the diversity of helicopter shapes allows each tree to buy hundreds of tickets in the aerodynamic lottery. Whether the sky howls, squalls, or puffs, the maples will have a helicopter design to suit the mood. The Taoist embrace of the wind is therefore a philosophy that applies throughout the tree's life. Leaves furl, trunks bend, and fruits are variable enough to conform to, then use, the wind's forceful nature.

May 18th—Herbivory

Springtime's perfect leaves have turned ragged. Their smoothness is broken by irregular gashes or tidily incised bite marks. The interminable storms of the past several weeks are partly responsible. A sassafras sapling hangs low, its leaves shredded by hail. Maple leaves are similarly slashed. This physical violence is dramatic, but it accounts for only a small portion of the damage borne by the mandala's leaves. The main culprits are the mouths of insects. They gnaw, suck, nibble, and rasp day by day, tearing down what the plants build up.

Half all insect species are plant eaters, and insects account for half to three-quarters of the species on earth. Plants are therefore plagued by six-legged robbers. Small plant species such as clover have to contend with one to two hundred species of herbivorous insects, while trees and other larger species have a thousand or more. These estimates are from northern areas, so the number of insect species that might browse or suck on each plant species in the mandala is likely much higher. Species richness is higher yet in the tropics. The world is full of marauding vegetarians; no plant escapes their attention.

The most obvious signs of herbivory in the mandala are holes in leaves. The bloodroot's leaves are naturally deeply indented, but insects have disrupted the flow of these lines with gouges and nips. Toadshade trillium likewise is etched with irregular gaps. Spicebush leaves are dotted with oval excisions, and perfect semicircles are carved out of its leaf edges. The perpetrators—or artists, depending on your perspective—

have left the scene. They are likely caterpillars, the larval stage of moths and butterflies. Caterpillars are champions of herbivory, designed to focus exclusively on turning leaves into insect flesh. But no caterpillars are visible, apart from one chewing on a maple leaf, the animal's pulsing gut visible through its thin green skin. I search leaf margins, stems, and growing tips, finding nothing. The insects are either hidden in the leaf litter or have moved up the food web, perhaps in the belly of a nestling bird.

Leaf miners have also left their mark, mostly in the leaves of seedling maples. Miners are like those humans who tear open a sandwich or cookie, eating the filling and leaving the crust. Miners do so not by opening the cookie but by diving inside, squirming their tiny flattened bodies between the upper and lower skins of the leaf. They tunnel into the cookie's center, munching at the cells inside, inching forward and leaving behind a feeding scar. Over one thousand species of miner work North America's leaves, and each species marks the leaf with its own style of scar. Some species move in circles, creating brown spots on leaves; others wriggle in seemingly random lines, scrawling thin paths across the leaf. More fastidious species move back and forth, systematically eating out the whole leaf, leaving a pattern like that in a freshly mown lawn. Leaf miners are the larvae of a taxonomically diverse selection of flying insects, including the young of flies, moths, and beetles. When the larvae have completed their work, they turn into winged adults that lay eggs on leaves, producing the next generation of miners.

The viburnum shrub in front of me has an entirely different kind of herbivore on its stem. The insect sits on the tender new growth at the tip of the shrub, perfectly color-matched in rich green. Its head is down, facing away from the stem's tip; its wings and body are slightly raised, shaped like an oriental slipper or a fancy Dutch clog. The overall effect is almost perfect mimicry of a bud. But this is no innocent bud. The green slipper is a leafhopper, an insect that attaches ticklike to its hosts.

Leafhopper jaws are stretched into a thin flexible needle that can wriggle between plant fibers, reaching down into the plant's blood vessels, the xylem and phloem. These are the same kinds of vessels that run up tree trunks, but in the thin-skinned new stems of viburnum the vessels are close to the surface and easily tapped by leafhoppers. Xylem carries mostly water, whereas phloem runs rich with sugars and other food molecules. Leafhoppers therefore prefer to feed on phloem, sliding their sharp mouthparts into the vessels. Because phloem is pressurized by the flow of sugary water from the leaves to the roots, leafhoppers simply tap into the vessels and let the plant squirt its food into their mouths. Leafhoppers, and their relatives the aphids, are so adept at tapping the phloem that they are put to use by scientists studying the plants. No human needle can match the exquisite delicacy of the insect's mouth, so researchers parasitize the parasite by snipping off the needle, killing the insect but leaving a probe lodged inside the phloem cells.

Insects feeding on plant sap face a bigger problem than the occasional sorry end in a lab. Phloem is a wonderful source of sugar but has few of the building blocks of proteins, amino acids. Xylem has little food of any kind. Phloem sap is ten to one hundred times poorer in nitrogen than are leaves, and leaves themselves are ten times poorer than animal flesh. Living on sap is therefore like trying to get a balanced meal from a case of soda. Leafhoppers solve the problem by drinking two hundred times their dry body weight in sap per day, equivalent to a human's drinking nearly one hundred cans of soda each day. This enormous volume compensates for sap's low concentration of nitrogen.

The leafhoppers' mega-drinking strategy creates another problem: how to flush out the excess water and sugar without also eliminating the nitrogen? Evolution has solved this problem by creating two paths for the phloem liquid that the leafhoppers drink. Their gut has a filter that sends unwanted water and sugar down a bypass, admitting only precious food molecules. The bypassed water and sugar is voided in drops from the anus, creating the sticky "honeydew" that coats plants infested with leafhoppers, aphids, or scales. Some entomologists claim that this honeydew is the manna that the Israelites ate during the Exodus. This is possible, of course, but it is hard to imagine anyone subsisting for forty years on the nutrient-poor excretions of leafhoppers, although honeydew supplemented by flocks of roasted quail might be feasible.

Even with a sophisticated filtering system in their guts, the diet of leafhoppers is inadequate, or it would be if they did not receive help from bacteria. Not only is plant sap watery but it contains an unbalanced mixture of amino acids; some of the amino acids necessary for insect growth are present, but some are not. Insects cannot make the missing amino acids from scratch. Instead, leafhopper guts have cells specially designed to hold bacteria that make amino acids. This is a mutually beneficial arrangement: the bacteria get a place to live and a continual supply of food, and the insects get their missing nutrients. Unlike the microbes that swim freely in the deer's rumen, these bacteria are embedded inside the cells of their host. Like the algae in lichens, the bacteria cannot live outside their host, nor can the host live without its internal helpers. The leafhopper on the branch in front of me is therefore a fusion of lives, another Russian doll in the mandala.

The dependence of leafhoppers on their bacterial helpers is of particular interest to entomologists in the pest control business. Leafhoppers and aphids exact a heavy toll on crops and often transmit disease to the plants that they puncture. If the relationship between the insect and its bacteria could be poisoned or otherwise disrupted, the entomologists might be able to wash fields clean of these troublemakers. This idea has yet to be put into practice, but I hope that if it ever is, we will not let the bright light of our ingenuity blind us to the possible costs of our actions. Chemicals that sever the tie between beneficial bacteria and their hosts may have effects well beyond ridding crops of leafhoppers. The soil's vitality depends on the action of such bacteria, as does the health of our own gut. At a deeper level, all animals, plants, fungi, and protists have ancient bacteria living inside their cells. Leafhoppers are the tip of the iceberg. Hammering at this tip risks sending fractures throughout.

The mandala contains insects designed to steal every part of a plant. Flowers, pollen, leaves, roots, sap are all preyed upon by a diverse toolbox of insect mouthparts. Yet the mandala is green. Leaves are a little tattered, but they still dominate the forest. Above, leaves are stacked in layers, blocking my view of the sky; around me, shrubs stretch out across the hillside, again hemming in my sight; below, my feet rest in a carpet of saplings and forest herbs. The forest seems to be an herbivore's heavenly banquet. Why is the mandala not stripped bare? This is a simple question, but it is much fought over, and it stirs up controversy among ecologists for good reason. The relationship between herbivores and plants sets the stage for the rest of the forest ecosystem. If we don't get the answer right, or if we cannot produce an answer, our understanding of forest ecology is shipwrecked and we are swimming in ignorance.

Birds, spiders, and other predators may give part of the answer. Their hunger might hack back the ravenous hordes of insects, protecting plants by preventing herbivore populations from getting large enough to fulfill their destructive potential. A corollary of this idea is that herbivorous insects seldom compete among themselves; they are suppressed by their predators, not by their peers. This is important because competition is the force that drives evolution. If herbivore populations were limited solely by predation, we would expect natural selection to have lavished more effort on helping herbivores to avoid predators than on giving them an edge in competition for food.

The idea that insect populations are suppressed by their predators has been tested by building cages around plants. If predation rules the insects' world, insect numbers should explode inside the cages, and caged plants should be eaten down to nubs and stumps. The results of caging experiments are mixed. Insect populations do sometimes increase when their predators are kept away, but the jump in numbers is seldom dramatic, and in some seasons and locations the cage has no effect at all. Even when cages do boost insect populations, the caged plants remain leafy green, albeit more chewed than their uncaged kin. Predation cannot, therefore, be the only explanation for the seeming paucity of herbivores.

We too are plant eaters, and our feeding behavior suggests another approach to the puzzle of the greenness of the woods. I live surrounded by maples, hickories, and oaks but have never sat down to a tree-leaf salad. Forest herbs grow in profusion at my feet but, again, I have not dined on them. My medicinal botany books tell me that small doses of the herbs in the mandala may alleviate ailments, but anything more than a nibble will cause (depending on the species of herb) heart stoppage, glaucoma, stomach upset, tunnel vision, or irritation of the mucous membranes. Our domesticated crops have had the toxins bred out of them, giving us a distorted view of the realities of herbivory. Granted, we have not evolved to be leaf eaters and lack the detoxifying biochemistry of most true herbivores, but our inability to eat most of the plants that surround us reveals an important point: the world is not as green as it seems. This point is reinforced by the very fact that other herbivores have specialized biochemical methods to neutralize the toxins in their food. The mandala is not a banquet waiting for guests to arrive but a devil's buffet of poisoned plates from which herbivores snatch the least deadly morsels.

Organic chemists confirm the experience of our taste buds. The world is a bitter place, full of deterrents, digestive disrupters, and poisons. Hawks know this also, using fresh greenery to line their nests and drive out fleas and lice. Consider also the *New York Times*. Insects grown in containers lined with old copies of the paper fail to reach maturity. The quality of the insects' reading material is not the culprit, although insects raised on the London *Times* mature into adults. The *New York Times* is printed on paper containing the pulped wood of balsam fir. The fir tree produces a chemical that mimics the hormones of its insect herbivores, thus protecting itself by stunting and neutering its enemies. The London *Times* is made from trees that lack hormonal defenses, making the pulped flattened remnants of their bodies safe to use as bedding for laboratory insects.

We can now turn our question around and ask not how plants manage to survive assaults by herbivores, but how herbivores cope with noxious plants. The puzzle is no longer that the world is green but that the greenery has holes made by creatures who do not die after dining. Detoxifying countermeasures are the foundation of the herbivores' ability to eat poisoned plants, but the insects also try to dodge around the defenses by feeding on those parts of plants they are most able to digest. It is no coincidence that the green caterpillar in the mandala is feeding on *young* maple leaves. Maples, like many tree species, defend their leaves with bitter tannins. Tannins are effective deterrents only in high concentrations, so young leaves have not yet accumulated enough of these chemicals to make them noxious. If the same caterpillar were to hatch in August it would face a forest steeped in tannin. The springtime emergence of many herbivores allows them to sidestep the plants' defenses.

The biochemical swordplay between plants and their herbivores has created a tense stalemate in the mandala. Neither side has yet routed the other. The holes and incisions of the mandala's leaves are the marks of this year's round of cut and parry, the venerable duel from which emerges the mandala's fundamental character.

May 25th—Ripples

Hungry ladies dance in the air, swoop at my arms and face, then land and probe. They have flown upwind, excited by my smelly mammalian promise. No doubt the nakedness of my skin further stimulates them; no dense mat of hair to obscure their dinner table. What an easy meal!

One of the mosquitoes lands on the back of my hand, and I let her probe my skin. She is mousy brown, just a little furry, with scallop patterns along her abdomen. Slender curved legs hold her body parallel to my skin. A needle juts from under her head. She slowly moves this lance across my skin, seeming to test for a suitable spot. She stops, holds steady, then I feel a burn as her head drops between her forelegs and the needle slides in. The sting continues as she penetrates deeper, sliding in several millimeters. The sheath that held the needle has bent back between her legs, leaving a tiny length of thin tube exposed between her head and my skin. The needle looks like a single shaft, but it is a bundle of several tools. Two sharp stylets help cut into the skin, making way for salivary tubes and a strawlike food canal. The salivary tubes ooze chemicals that prevent blood from clotting. These same chemicals cause the allergic reaction that we call a mosquito bite.

The needle is flexible, so it bends after it enters the skin and, like a worm seeking a soft patch of soil, it probes around inside my skin, sniffing out a blood vessel. Capillaries are too small, so the mosquito searches for a larger vessel, a venule or arteriole, the state highways of our blood system. Veins and arteries, the interstates, are too tough-coated to be of interest. When the needle finds the object of its search, the sharp end punctures the vessel's wall. The flow of blood over the needle stimulates nerve endings that signal pumps in the insect's head to start pulling blood. If the mosquito fails to locate a suitable vessel she will either pull out and try again, or feed on the small pool of blood that the needle created as it tore through capillaries in the skin. This pool-feeding method is much slower, so most mosquitoes that fail to hit a good-sized vessel prefer to pull out and try again, seeking a rich seam elsewhere under the skin.

The mosquito on my hand has evidently pierced a productive vessel. In just a few seconds her light brown underbelly balloons into a shining ruby. The brown scallops on her back that mark each segment of the abdomen move apart, seeming to dislocate her tidy body. She rotates as she feeds, perhaps pushing the needle around a curve in the blood vessel. When her belly is stretched into a half globe, she abruptly lifts her head and, in a blink, flies off. I am left with a slight burn on my hand and two milligrams less blood.

These milligrams are a trifle for me, but they have doubled the mosquito's body weight, making her flight ponderous. The first thing she'll do after finishing her meal is rest on a tree trunk and offload by urination some of the water she has swallowed. Human blood is much saltier than a mosquito body, so she'll also pump salts into the urine, preventing my blood from disrupting her physiological equilibrium. Within an hour she will have dumped about half the water and salt from her meal. What remains, the blood cells, will be digested, and my proteins will find themselves turned into yolk in a batch of mosquito eggs. The mosquito will also take some of my nutrients for herself, but the vast majority will be used for egg production. The millions of mosquito bites inflicted on us every year are therefore preliminaries to mosquito motherhood. Our blood is their ticket to fecundity. Male mosquitoes and females who are not breeding feed like bees or butterflies, sipping nectar from flowers or drinking sugars from rotting fruit. Blood is a proteinaceous boost for mothers only.

The mosquito's colors and fuzz mark her as a member of the *Culex* genus. This means that she will lay a small raft of eggs on the surface of a pond, ditch, or stagnant pool. *Culex* often breed in the fetid water that surrounds human habitations, giving them their common name, "house mosquito." Females fly a mile or more from these breeding areas, searching for a suitable blood donor. My blood may end up in eggs laid in the pond half a mile behind me or in a blocked gutter or sewer in the town a mile away. Here the eggs will hatch into aquatic larvae that live suspended just under the water's surface. Their rear end is an air tube that adheres to the water's surface film, providing both an anchor and a breathing hole. Their heads hang down into the water and filter bacteria and dead plant matter from the cloudy water. During their life cycle, mosquitoes therefore exploit three of the richest food sources available to animals: the bounty of wetlands, the concentrated sugars in nectar, and the viscous feast of vertebrate blood. Each meal propels them into their next life stage, creating a nearly unstoppable momentum.

If I had not visited the mandala the *Culex* would likely have found another blood donor for her meal. Despite their fondness for human habitations, *Culex* usually feed on bird blood. This is to the birds' detriment because *Culex* transmits disease, most notably avian malaria and, of late, West Nile virus. Avian malaria lives in the blood of about one-third of the birds flying above the mandala. Most of the infected birds appear to be able to live out their lives without being substantially weakened by the parasite. Birds infected with the invading West Nile virus have higher mortality, perhaps because America's birds have no natural resistance to this virus from Africa.

When *Culex* mosquitoes cannot find a crow or chickadee, they feed on humans. This flexibility of dining arrangements brings bird parasites into contact with human blood. Some, like avian malaria, die in the alien surroundings. But others, including West Nile virus, will sometimes take hold and infect the human. This jump from bird to human blood requires that a mosquito first feed on an infected bird and pick up the virus, which then multiplies in the mosquito's salivary glands. If the mosquito then feeds on a human, the drop of saliva may now carry an unwelcome guest, and the West Nile virus may jump from crow to human.

Perhaps I should not have been so sanguine about my exsanguination; my curiosity may have allowed another life-form to take over my body, perhaps even kill me. I am, however, hardly dancing on the precipice. In the whole of North America only four thousand people were infected with West Nile virus last year, fifty-six of them in Tennessee. About fifteen percent of these cases are fatal, making the virus fearsome if you get it but, compared to the other risks we face each day, a very minor threat. The newsworthiness of the virus comes not from the magnitude of the threat that it actually poses to us but from its novelty, its indiscriminate choice of targets, and our inability to predict whether it will bloom into a larger threat. The virus is also a gift to pesticide manufacturers, scientists feeding from the government's largesse, and news editors desperate for sensational copy. Fear and profitability have launched the virus to stardom.

Until recently a much more deadly threat to humans hung over the mandala. Another species of malaria lurked in mosquito salivary glands, waiting not for a bird but for a human. In the first years of the twentieth century the average rate of mortality from malaria for people living in the southern United States was about one percent per year. In the swamplands of Mississippi the rate was three percent; in these Tennessee hills it was lower but still significant. Malaria's terrible weight used to oppress people all across the eastern United States, but eradication programs eliminated it from the Northeast in the nineteenth century, decades before clearing the South. Malaria's end in the South came about in the early twentieth century after a campaign that targeted many stages of the malarial life cycle. Huge quantities of quinine were distributed to treat infected people and prevent reinfection of mosquitoes. Screens on windows and doors were encouraged or required, breaking the link between mosquito saliva and human blood. Wetlands and ponds were drained to remove breeding sites for mosquitoes, or oiled to smother their larvae, or poisoned with insecticides. Although malaria's hosts, mosquitoes and humans, still lived across the South, the distance between them was stretched sufficiently that the parasite dropped into extinction.

Malaria is seemingly irrelevant to my modern experience in the mandala, but this is an illusion. The mandala has been spared the chainsaw because it lies in an area set aside by the University of the South. This university brought me here also. What brought the university to this hillside? Malaria, among other things. Like many of the older universities in the East, the school is located on a plateau, away from the swamps that breed malaria and yellow fever. The cool temperatures and relative freedom of the Tennessee hills from mosquitoes made them an ideal place to send the offspring of the southern gentry. The school year ran through the summer, allowing students to escape the heat and disease of the cities. School was closed and abandoned in the winter, when the mosquitoes of Atlanta, New Orleans, and Birmingham were in abeyance. This prime location helped cement the university to the mountaintop, ensuring its viability long after one of its primary benefactors, the malarial parasite, had faded from the land.

The atoms that make up my blood were propelled to the mandala by these biological forces of history, and it is appropriate that a mosquito should carry away some of them to rearrange them into a raft of eggs. This physical connection to the rest of nature is often unseen. The mosquito bite, the breath, the mouthful are acts that create a community, that keep us welded into existence, but that mostly pass unacknowledged. A few people say grace at a meal, but no one does so with every inhalation or insect bite. This unconsciousness is partly self-defense. The connections through the millions of molecules we eat or breathe or lose to mosquitoes are too many, too multifariously complex for us to attempt comprehension.

. . .

The whining reminders of interconnectedness persecute me as I sit at the mandala, so I pull up the hood of my sweatshirt and tuck my hands into the sleeves, trying to slow the barrage. I peer over the lip of my cocoon and study the evidence of another kind of atomic flow. A snail has been smashed on the rock beside me. A few translucent crumbs of the honey-colored shell lie on the surface of the rock. These are the remains of a bird's calcium-hungry feast.

The crushed snail in the mandala is one stream among many in the great springtime flow of calcium from the soil to the air. Breeding female birds scour the forest for snails, greedy for the sheets of calcium carbonate on the snails' backs. Such hunger is well founded. Without a boost of dietary calcium the birds cannot make their chalky eggshells.

Once a snail has been swallowed by a bird, the shell is first ground up in the bird's gizzard, crushed by a knot of muscle and pieces of hard sand. The calcium then gradually dissolves into the mushy gut and is pumped across the walls of the intestines into the bloodstream. If the bird is laying eggs that day, the calcium may go straight to the reproductive organs. If not, it will go to special calcium storage areas in the core of the long bones of the bird's wings and legs. This "medullary bone" is produced only in sexually active females. Over the course of a few weeks the medullary bone is built up in preparation for egg laying, then completely disassembled as the eggs are laid. Female birds take to heart Thoreau's wish to "suck out all the marrow of life"—they suck dry their own bones to make new life each spring.

The sucked calcium travels through the blood to the shell gland. Here the calcium carbonate leaves the blood and is added in layers to the egg. The shell gland is the last stop through the tube that carries the egg from a bird's ovaries to the outside world. The earlier stages of this journey wrapped the egg in albumen, then two layers of tough membrane. The outermost membrane is studded with tiny pimples that bristle with complex proteins and sugar molecules. These attract calcium carbonate crystals in the shell gland and act as centers from which the crystals can grow. Like sprawling cities, the crystals build on one another and eventually join, creating a mosaic across the surface of the egg. In a few places the crystals fail to meet, leaving an untiled hole in the mosaic that will become a breathing pore extending from this first layer of the eggshell all the way to the surface of the completed shell. The next layer of calcium carbonate grows on top of the first, creating a shell made from pillars of calcium crystals pressed closely together. Protein strands weave across these pillars, adding reinforcement to the shell. When the thickest layer of the shell is complete, the shell gland lays a pavement of flat crystals over the shell surface and then paints the pavement with a final protective layer of protein. Thus has the snail's shell been uncoiled, rebuilt into an avian cocoon.

As the young bird grows inside the egg it pulls calcium out of the shell, gradually etching away at the walls of its home, and turns the calcium into bone. These bones may fly to South America and be deposited in the soil of the rain forest, or the calcium may return to the sea in a migrant-killing autumn storm. Or, the bones may fly back to these forests next spring and, when the bird lays her eggs, the calcium may again be used in an eggshell whose remains may be grazed on by snails, returning the calcium to the mandala. These journeys will weave in and out of other lives, knitting together the multidimensional cloth of life. My blood may join the snail's shell in a young bird that eats or is bitten by a passing mosquito, or we may meet later, in millennia, at the bottom of the ocean in a crab's claw or the gut of a worm.

Winds of human technology blow at this cloth, billowing it in unpredictable directions. Atoms of sulfur that were locked into fossil plants when they died in ancient swamps are now tossed into the atmosphere when we burn coal to fuel our culture. The sulfur turns to sulfuric acid, rains down on the mandala, and acidifies the soil. This acidic fossil rain tips the chemical balance against the snails, reducing their abundance. Mother birds have a harder time bingeing on calcium and so breed less successfully, or not at all. Perhaps fewer birds will mean less blood for mosquitoes, or fewer predatory beaks? Viruses like West Nile that thrive in wild birds may, in turn, be touched by the changed bird populations. This ripple in the cloth floats across the forest, perhaps finding a hem at which to end, perhaps floating on forever, drifting through the mosquitoes, viruses, humans, ever outward.

June 2nd—Quest

A tick perches at the tip of a viburnum branch, a few inches from my knee. I suppress the urge to flick the pest away. Instead, I lean in to see the tick for its own sake, trying to look beyond my quick mental dismissal of it as a mere pest. The tick senses my approach and lifts the front four of its eight legs in a frenzied wave, grasping at the air. I wait, still, breath held, and the tick relaxes back into its original posture with just its front pair of legs raised in a prophetlike salute to the sky. My eye is so close that I see tiny scalloped ornamentations around the edge of the tick's leathery oval body. The raised legs have translucent feet at their ends, each of which catches the sun and glows. In the center of the back is a white spot, identifying the animal as an adult female lone star tick. The chestnut color from the rest of the body seems to bleed into the star, giving it a golden sheen.

Ugly, unadorned weaponry on the tick's head counterbalances the strange beauty of the rest of her body. The head is tiny, unnaturally so, and through my hand lens I see two stubby pillars jutting forward, barely covering a Swiss Army knife of sharp, grotesque mouthparts. I want a closer look at this nastiness, so I reach up, hold the viburnum, and pull it toward my eye. The tick senses my hand and snaps toward it, semaphoring wildly with her forelegs. This sudden catapult startles me and I jerk my hand back, releasing the branch, sorely disappointing the tick.

This foot-waving tick in the mandala is engaged in what zoologists call questing behavior. This gives the animals a measure of Arthurian nobility, tempering our disgust at their bloodsucking habits. The image of a quest is particularly apt because both the Knights of the Round Table and the Arachnids of the Leafy Forest seek the same end: a blood-filled Grail. In the case of the lone star tick, this Grail is a warm-blooded animal, either a bird or a mammal.

The knights' mythical quest led them to the blood from Christ's wounds, collected in the Grail by Joseph of Arimathea. The ticks are less selective about the theological pedigree of the blood they seek, and their quest ends with molting or sex. The ticks' quest also differs substantially in style from the journeys of the knights. Most ticks sit and wait for the Grail to come to them, then ambush it, rather than trekking across continents to hunt down their meal of blood. The tick in the mandala showed the classic approach to questing: climb up a shrub or blade of grass, position yourself at its tip, then hold out your forelimbs and wait for your victim to brush up against you.

The ticks' quests are aided by Haller's organs on each foreleg. These spiny indentations are packed with sensors and nerves, tuned to pick up a waft of carbon dioxide, a puff of sweat, a tiny pulse of heat, or the thudding vibrations of footsteps. The raised forelegs therefore serve both as radars and as graspers. No bird or mammal can pass near a tick without being detected by smell, touch, and temperature. When I pulled the viburnum branch and breathed on the tick, I sent its Haller's organs into a jangling spasm, unleashing the tick's springlike lunge at my finger.

Dehydration is the ticks' main foe during their quests. Ticks sit in exposed locations for days, even weeks, waiting for their hosts. The wind whisks away moisture, and the sun bakes their small leathery bodies. Wandering off in search of a drink would interrupt the quest and, in many habitats, there is no water to be found. So, ticks have evolved the ability to drink water from air. They secrete a special saliva into a groove near the mouth and, like the silica gel that we use to dry our electronic gadgets, their saliva draws water out of the air. The ticks then swallow the saliva, rehydrating themselves and continuing the quest.

The quest ends when the forelegs lock on to the skin, feathers, or hair of a potential host. The lucky tick then crawls over the host, testing the skin with its mouthparts, probing for a soft, bloody site to attack. Like cat burglars, the ticks scramble over our bodies without raising the alarm. Take a pencil and run it lightly up your arm or leg. You'll feel it. Take a tick and let it crawl over your limbs. More than likely you'll not feel a thing. No one knows how they do this, but I suspect that they charm our nerve endings, taming the cobralike neurons with the hypnotic music of their feet. The best way to detect a tick crawling up your leg is to notice a suspicious absence of tickling and itching. Walking in the forest in summer generates an endless stream of entomological creepies on your skin. When the stream of sensation dries, you've got a tick.

Unlike mosquitoes, ticks take their time about feeding. They press their mouthparts against the skin, then slowly saw into the flesh. Once this inelegant incision has opened a large enough hole in the skin, they lower a barbed tube, the hypostome, to draw out blood. A full meal takes days to extract, so ticks cement themselves into the skin to prevent the host from scratching them off. The cement is stronger than the tick's own muscles, explaining why burning ticks with matches is futile. Ticks cannot pull out rapidly, even when their rear ends are on fire. Lone star ticks feed deeper than other species, making them particularly hard to remove.

The blood meal causes ticks to swell up so much that they grow new skin to accommodate their meal. They drink so much blood that they face a reversal of the dehydration problem of their questing days. Rather than curtail their meal when they are full, the ticks extract water from the blood in their gut, then spit it back into the host, an action that surely violates the spirit, if not the letter, of the laws of chivalry, particularly if the tick carries one of the many bacteria that cause disease. The half teaspoon of blood in a fully engorged tick is therefore a distillation of several teaspoons of host blood, thickened and stored in the tick's belly.

A feeding female adult tick will increase her body weight one hundred times, then summon her lovers from elsewhere on the host. She releases pheromones while still attached, and these airborne chemicals create a scramble of males toward her voluptuousness. Once a male arrives, the female releases more pheromones, and the male crawls under the swollen enormity of his immobile mate. He uses his mouthparts to insert a small package of sperm into a chink in the female's armor and then leaves her to finish her meal. When she is fully sated, she dissolves the cement around her mouth and crawls, or drops, onto the ground. There, she slowly digests the blood and fills thousands of eggs with nutritious yolk. Like the mosquito, the mother tick uses blood to fuel her reproduction. When the eggs are ready, she lays them in clusters on the forest floor. Her quest is over, the Grail's blood has been transubstantiated into the Body of Tick Egg, and she dies empty but fulfilled.

A week later the dreaded "seed ticks" emerge from the eggs. These larvae look and act like miniature versions of their parents as they swarm up the vegetation around their hatch site to begin their quest. Because they emerge in clusters, they attack hosts en masse, multiplying our misery. Only one in ten of these larvae succeeds in finding a host. Most starve or dry out before a suitable animal passes by. Lone star larvae attack birds, reptiles, and mammals except for rodents, which they seem to avoid. The larvae of other tick species reverse this preference and seek out mice and rats for their first meal. Successful larvae feed in the same way as adults, then drop off and molt into a slightly larger form called a nymph. The nymphs quest and feed, then molt into adults. The adult tick in the mandala has therefore already completed two successful quests. She may be two or three years old, having overwintered as a larva and then again as a nymph.

I am tempted to repeat my experiment with the mosquito and reward this tick's longevity with a gift of my blood. I pass up the opportunity for two reasons. First, my immune system reacts violently to tick bites, leaving me itching and, if I have more than a few bites, sleepless. Second, unlike with the mosquito, there is a fair chance that this tick is carrying a nasty disease. The most famous tick-borne malady, Lyme disease, is fairly rare here and is seldom carried by lone star ticks. However, lone star ticks are the primary carriers of other diseases including Ehrlichiosis and the mysterious "southern tick-associated rash illness." These latter bacteria have yet to be propagated outside the human body, so we know very little about them, other than they cause a Lyme-like illness. Rocky Mountain spotted fever and malaria-like babesiosis also may lurk inside this lone star. This bestiary of pathogens is a powerful disincentive to offer myself.

Despite the nobility of the tick's quest and my admiration for her armor and weaponry, I feel a strong need to flick her away or pinch her with my fingernails. Such disgust may come from somewhere deeper than mere learned precaution. Fear of ticks is etched in my nervous system by the experience of many, many lifetimes. Our battle with questing ticks is at least sixty thousand times older than the Arthurian legends. We've been scratching and tweaking at ticks through our entire history as *Homo sapiens*, back to our early primate days when we chattered and groomed one another, to our time spent as itchy insectivores and, further, to our reptilian origins when ticks evolved ninety million years ago. The Grail tires after so many millions of years of pursuit. I skirt the viburnum bush as I leave.

June 10th—Ferns

We are on the cusp of summer. For the last two weeks, the temperature and humidity have edged higher day by day, and the heat has slowed my walk to the mandala; the time for the vigorous, warming hikes of winter has long passed. In the forest, the abundance of animal life is impressive, especially when contrasted with the quiet of winter. Birdsong streams from all directions. The air is busy with the continual passage of small midges, mosquitoes, wasps, and bees. Ants crawl over the litter, several dozen visible at any moment within the mandala's circle. Furry jumping spiders also ply the forest floor, and millipedes trundle through gaps in the litter. Above, the tree canopy is dense and multilayered. The leaves have matured from the light, airy green of spring to summer's more substantial, deeper hues. Photosynthesis in this thick canopy of leaves is now at full capacity, harvesting the energy that forms the foundation of the forest ecosystem.

At ground level, the spring ephemeral wildflowers have mostly faded away. The plants that are left are all shade specialists, growing slowly in the dim understory. Of these species, the ferns are the most abundant and conspicuous. As I look across the forest floor, I see ferns emerging every meter or so across the whole mountainside.

Christmas fern fronds as long as my forearm arch like jaunty hat plumes on the southern edge of the mandala. Their fresh growth reaches over last year's fronds, which, although they are still attached to the fern's base, lie prostrate and dying. The old fronds stayed green through the winter and spring, giving the plant a photosynthetic boost before the emergence of this year's new growth. The hardiness of the Christmas fern gave European settlers greenery with which to decorate their winter festivals, and the fern was named to honor this use. The new growth on these plants emerged in the mandala in April, pushing through the litter as silvery, tightly coiled fronds called fiddleheads. As the coils unfurled, their central stems lengthened and leaflets grew, producing elegant, tapered feathers.

The leaflets at the tips of the highest-reaching fronds are pinched and shrunken. Instead of presenting the sun with a broad surface for photosynthesis, these attenuated leaflets carry two rows of discs on their undersides. The discs are about as wide as peppercorns. Like a skullcap pressed onto a head of curls, each disc has a mass of brown fuzz protruding from its edges. When I peer through the hand lens, the mass of curls is transformed into a carpet of dark snakes. Each snake's body is divided into sand-colored segments with wide mahogany edges. The snakes' mouths hold fat clusters of golden globes. I see no movement today, but previously I have seen the snakes rear up, then snap back, spitting the globes high into the air.

The globes are the fern's spores, each one with the makings of a new fern enclosed in its tough coat. The snakes are botanical catapults, designed to hurl spores skyward. The snakes' segments are cells with unevenly thickened walls, and this unevenness powers the motion. On sunny days, the water that lines the cells evaporates, increasing surface tension in the remaining water. Because the cells are so small, the increasing surface tension is strong enough to bend them, arching the snake upward. As the snake rears, it scoops up a mass of spores, preparing to launch them. More water evaporates and the tension increases, bending the snake further. Snap! The tension breaks and the cell walls' pent-up energy flicks the spores away. When the sun shines directly on a ripe frond, water evaporates rapidly from the snakes' cells, sending spores flying like popcorn from hot oil. To the naked eye these escaping spores look like puffs of smoke. Under a lens, the action is more dramatic: the barrage of snapping catapults looks like a battlefield reenactment.

The catapults' dependence on the sun's drying power allows the ferns to launch spores only on dry days, when the spores have a good chance of traveling far. Today, the air is thick with humidity, the sky is darkening gray, and thunder growls in the distance. This is an inauspicious time for traveling spores; they risk being washed out of the air, and so the catapults lie motionless.

Like an animal's egg or sperm cell, each spore carries a shuffled deck of exactly half its parent's genes. But unlike an egg or sperm, a spore will land, then germinate without uniting with another spore. This is the first hint that the life cycle of plants is radically different from our own. Sex for animals is a quick two-step: make the sex cells by halving your genetic library, then fuse egg and sperm into a new animal. Just two stages, a simple rotation. But ferns are up to something strange. No frond emerges when the spore germinates. Instead, a small "lily pad" grows, spreading out its flat body until it is the size of a small coin.

The lily-pad fern makes its own food and lives as a separate individual. After a few months or years, swellings appear on its skin. Some swellings are like blisters; others are like tiny chimneys. The blisters expand, then on a rainy day they burst and release sperm cells. These spin through surface water, sniffing for chemicals from an egg sitting at the base of a chimney. Each chimney's core is filled with chemicals that bind and destroy sperm cells of the wrong species. Suitable sperm face no such bar and swim to the egg. The two cells then fuse. The resulting embryo develops into a new Christmas fern that will eventually grow large enough to fling spores from the tips of its arching fronds. The fern's life cycle therefore has four steps: spore, lily pad, egg or sperm, and large fern.

On the other side of the mandala a rattlesnake fern adds some interesting twists to this life cycle. Its leaves spread low over the litter in a lacy fan that is about as wide as the span of my hand. A spike rises from the center of the fan, reaching twice as high as the leaves. At the top of the spike, several dozen millimeter-wide capsules cluster on small side branches. These capsules shake spores from vertical slits in their sides. The spores germinate and grow not into lily pads but into subterranean tubers, like miniature potatoes. These tubers have no chlorophyll and depend on a fungus for food. After several years of growth, the tuber makes sperm and eggs, which will then produce another rattlesnake fern.

Once grown, the adult rattlesnake fern continues to exchange nutrients with the fungus. Some rattlesnake ferns take this mutualistic relationship to its extreme, never lifting their fronds above the leaf litter. These individuals grow and produce spores entirely belowground, nourished by their fungal collaborators.

Both species of fern in the mandala alternate between two lifeforms: the large spore-bearing plant and a smaller egg-and-sperm factory, the lily pad or tuber. This alternation between different identities is hard for humans to grasp, and the sex life of ferns remained a mystery until the 1850s. The obvious reproductive structures, the spores that blew away in the wind like pollen or seeds, were unlike any other sex cell. Botanists called ferns and their similarly confusing kin, the mosses, cryptogams or "hidden sex" plants, applying a terminological Band-Aid to the bothersome mystery. The confusion lifted once sperm and egg cells were discovered swimming in the watery surfaces of tiny lily pads.

The fern's reproductive methods are well suited to life in protected, wet places, but ferns fare poorly under drier, more rigorous conditions. Without moisture for their sperm to swim in, ferns cannot reproduce. In addition, their lily-pad stage offers little protection or nourishment for embryos. Flowering plants have broken free from these constraints by modifying the fern's life cycle. Instead of releasing spores into the wind, flowering plants produce spores that are retained within the flower's tissues. These spores grow into protected miniature lily pads that then make eggs and sperm. Thus the fern's independent lily pad was shrunk to a few cells buried inside the flower. This freed flowering plants from the need to find a watery nook in which to reproduce. Deserts, rocky ridges, and dry hillsides were no longer barriers to plant sex, nor were dry spells or sunny days. Whittling down and retaining the lily pad also allowed flowers to nurture their offspring, passing nourishment to them, wrapping them in protective seed coats, and holding them aloft inside fruits to catch the wind or be picked up by a passing seed-dispersing bird.

The reproductive innovations of flowering plants allowed them to become by far the most diverse group of plants alive today. There are over a quarter of a million species of flowering plant but just over ten thousand fern species. When flowering plants evolved, about one hundred million years ago, many species of ancient ferns and other nonflowering plants were displaced, outcompeted by the newcomers. It would be a mistake, however, to dismiss modern ferns as primitive leftovers. Recent studies of DNA show that modern ferns evolved and diversified *after* the rise of the flowering plants. As the flowering plants took over, they caused ancient ferns to disappear, but they also inadvertently produced the ideal conditions for a new dynasty of ferns: moist mandalas in which shade-loving Christmas and rattlesnake ferns can thrive.

June 20th—A Tangle

A gusty wind is clearing the sky of the clouds that have drizzled rain all week. The first sunshine in days is breaking through small gaps in the tree canopy, checkering the mandala with patches of light and shadow. The slick surfaces of *Hepatica* leaves flash as the sunlight catches them. Other plant species lack *Hepatica*'s shine but glow with various shades of green. After so many days under dull skies, the mandala's colors seem particularly vivid. The forest also sounds more alive. From all around me comes a gentle hum, the combined buzzes of thousands of insect wings, like the sound of a distant beehive.

Although it is midmorning and the sun has been out for hours, two snails lie exposed on the damp leaf litter. They have likely been here since before sunrise, twisted together in a mating tangle. Their horn-colored shells face each other, aperture to aperture, and their bodies are fused in a knot of gray-and-white flesh. These snails are locked in a difficult negotiation and exchange. Instead of transferring sperm from male to female, as most animals do, the snails are moving sperm both ways. Each individual is both a donor and a recipient of sperm, male and female united in one body.

Hermaphroditism creates a complex economic problem: how to make sure that the reproductive exchange between partners is fair. For snails, as for most organisms, sperm is cheap to produce but eggs are expensive. In unisexual animals these different costs generally favor selectivity on the part of females and unselective promiscuity on the part of males, especially in species where males contribute nothing to raising the young. In hermaphrodites, though, discernment and promiscuity are joined in one body, and mating becomes fraught, with each individual being cautious about receiving sperm while simultaneously trying to inseminate its partner.

Snails that detect a whiff of disease on their mates will refuse to express their female side, giving but not receiving sperm. Snails that find uninfected mates, however, will readily accept sperm from their partners. This selectivity may help snails pick out genetically superior sperm for their limited supply of eggs. Hermaphrodites are also sensitive to their larger social context. When they live in an area with few potential mates, they express both their male and female sides, but in a more crowded situation their femininity is dimmed and they act more like males, freely giving sperm but saving their eggs for only the best partners. The situation is yet more complex if mates have already bred and accepted sperm from another individual. One partner may then refuse another altogether, causing the rejected partner to attempt to mate forcibly, shoving sperm packets at the unwilling partner. Love triangles are fraught; love hexagons are war zones.

War is no metaphor. In some species, snail mating tensions have ballooned into armed conflict: bony darts shoot from one mate to another, sperm-destroying glands neutralize unwanted maleness, and muscles jostle sperm and eggs into battle lines. Even the length of the snails' embrace may be a result of sexual conflict. The snails palpate each other with their tentacles, circle, slowly edge into position, always ready to pull back or realign. What the snails are assessing at each stage is unknown, but their extended courtship and copulation is choreographed like cautious diplomacy, a prenuptial conference over the terms of the union. Such languor must surely be costly. The snails in the mandala have lain with their bodies mostly unshelled for over half an hour, easy prey for birds or other predators.

Hermaphroditism is an unusual sexual system for animals, most of which divide male and female functions into separate bodies. Yet all land snails are hermaphrodites, as are some sea-dwelling mollusks and a scattering of other invertebrates. The sexuality of the snails in the mandala has more in common with springtime flowers than it does with the birds or the bees. All the mandala's spring ephemerals and trees are hermaphrodites, and many of them combine male and female in a single flower. This diversity of sexual systems is puzzling. Why should wrens be boys and girls, but the trees that the wrens live in be boy-girls? The beetles that the wrens feed to their nestlings are either male or female, but the snails that arrive in the same beakful of food are all hermaphrodites.

Evolutionary theorists have treated this puzzle as a problem in natural economics. Just as a business manager decides how best to allocate the company's resources, biologists conceptualize natural selection as a process that decides how organisms will invest their reproductive energies. The human manager uses foresight and reason, but natural selection works by continually throwing out new ideas, then weeding out the ineffective in favor of the fecund. Nature has no shortage of new sexual ideas: every generation of snails has a few individuals that are unisexual, just as a small number of birds, insects, and mammals are born as hermaphrodites. There is therefore plenty of raw material to stimulate nature's free market of sexual roles.

Every individual has a limited supply of energy, time, and flesh to devote to breeding. Organisms can act like specialist companies and invest their resources in just one sex, or they can diversify and split the investment into two separate ventures, male and female. Which strategy is best depends on the particularities of each species' ecology. In situations where individuals face a high probability of not finding mates, it pays to be a hermaphrodite. Tapeworms that live by themselves in a gut have to self-fertilize or their genetic lineage will end. Less obviously, flowers that use unreliable pollinators to achieve sexual union may also need to self-fertilize. *Hepatica* plants bloom in profusion all over the mandala, but if the springtime weather is too cold and pollinating insects cannot fly, being a hermaphrodite is the only way to reproduce. The same is true of weedy species that colonize disturbed ground. Individuals from these species may find themselves as the only immigrant in a new piece of habitat, so self-love is essential. Hermaphroditism is therefore the favored sexual system of species that may have to breed without mating.

But many hermaphrodites, including most snails, do not live isolated lives and cannot self-fertilize even if they are put into solitary confinement. Loneliness is therefore not the only cause of hermaphroditism. Evolution has also favored hermaphroditism when a generalist approach to sexuality is most fruitful. Snails defend no breeding territories and they produce no songs or colorful displays. They also provide no parental care for the eggs, laying them in shallow pits in the leaf litter, then abandoning them. This relative simplicity of reproductive duties allows snails to be simultaneously male and female without compromising the efficacy of either sex. This is not possible for species like birds and mammals with more specialized sexual roles. In these cases, natural selection favors focusing on either maleness or femaleness. In economic terms, a snail gets a better return from a mixed investment strategy that combines male and female, whereas a bird gets a better return by channeling all its investment into one sex.

The varied ecological and physiological context of each species in the mandala has produced, through years of natural selection, a wide variety of sexual arrangements. The snails' hermaphroditic embrace, seemingly so alien to most humans, is a reminder that sexuality in nature is more malleable and diverse than we might at first suppose.

July 2nd—Fungi

Rain has gushed over the mandala for two days and two nights. The storm blew in from the Gulf of Mexico, and its incessant pounding has cleared the air of biting insects, bringing relief from the throngs of mosquitoes that have been my enthusiastic companions for weeks. The hottest days of summer have arrived on the heels of the storm. The air's fevered humidity now has a relentless, all-encompassing quality; any bodily exertion brings a sheen of sweat. The forest is held in a clammy tropical embrace.

Specks of orange, red, and yellow, the sexual buds of fungi, glow from the sodden forest floor. The heat and rain have emboldened the belowground parts of fungi, causing them to sprout their fruiting bodies. The prettiest of this morning's colorful fungi is a cup fungus perched on a decaying twig. Tangerine orange, shaped like a goblet, and fringed with silver hairs, it is called a shaggy scarlet cup. Although it measures less than an inch across, its color catches my attention, drawing me onto my knees to examine it more closely. Once my eyes are closer to the ground, I see tiny fruiting bodies everywhere, a colorful regatta on a sea of decaying leaves and twigs.

These bright boats all belong to the largest division of the fungal kingdom, the sac fungi, named for the sacs in which their spores are grown. The shaggy scarlet cup in the mandala started life as a spore, measuring just two-hundredths of a millimeter across, blown onto the dead twig on which it now lives. The spore germinated, then grew a slender filament into the twig's wood. Because fungal filaments are so skinny, they can slip between plant cell walls and snake through the minute pores between the cells. Once inside the twig, the growing filament oozed digestive juices, liquefying the seemingly tough wood. The fungus filament absorbed sugars and other nutrients from this disassembled wood soup, building new filaments that slid farther into the twig's dead tissues. To be locked in a wooden box below the ground is the shaggy scarlet cup's delight.

Some of the other participants in today's regatta also specialize in deconstructing twigs, whereas others prefer mats of dead leaves. But despite differences in taste, all these fungi grow the same way, by creeping their tentacles through dead plant material, enlarging their networklike bodies by feeding on, and ultimately destroying, their wooden surroundings. When fungi feed, they push their homes deeper into the sea of oblivion. Dead twigs are therefore sinking islands of habitat, and fungi must continually send out progeny to seek new islands. It is this imperative that brings the fungi into our sensory world. Fungi remain hidden from our eyes until their belowground filaments sprout fruiting bodies. The flotilla of yellow, orange, and red is a reminder of the vast network of life below the mandala's surface.

The shaggy scarlet cup produces propagules on the inner surface of its cup. Here lie millions of cannon-shaped sacs, each pointed at the sky with eight minute spores loaded inside. When the cannons are ripe, their tips snap off and the spores are fired into the air, gunning several inches above the cup and escaping the calm boundary layer of air that hugs the mandala's surface. Each spore is so small that it is invisible to the naked eye, but the simultaneous eruption of millions of spores looks like a puff of fine smoke. A cup's eruption can be triggered by a gentle touch to just one part of the cup's surface. This makes me suspect that animals might be important dispersers of fungal spores, despite the textbooks' assertions about "wind dispersal." This morning, the mandala's surface has within its circumference at least eight millipedes and centipedes (one of which is nibbling an old shaggy cup), several spiders, a large beetle, a snail, several dozen ants, and a nematode. Squirrels, chipmunks, and birds hop around the mandala's edges. Sac fungi fruiting bodies are so densely packed on the surface that it would be hard for animals not to step on them, even if they tried.

A small brown mushroom in the center of the mandala showers spores from open gills, instead of shooting them up from the ground as do the sac fungi. Wind is again believed to be the primary carrier of these spores, but animals have left their mark here too. The mushroom's cap is untidily scalloped with bite marks, perhaps from a chipmunk whose nose and whiskers are now brushing spores onto leaves many meters away.

The reproductive life of sac fungi and mushrooms is without parallel in the living world. They stretch the meaning of "sex" beyond anything we animals have achieved even in our most innovative moments. They have no separate sexes, at least none that we would recognize, and they do not make sperm or eggs. Instead, fungi reproduce by merging their filaments, literally melding their bodies to make the new generation.

The mushroom in the center of the mandala gives the clearest view into this strange life cycle. When mushroom spores germinate, they produce baby filaments that grow through the dead leaves, seeking mates. Filaments exist not as male or female but as different "mating types." These mating types all look the same to us, but fungi use chemical signals to sense the differences and will reproduce only with a mating type that differs from their own. Some fungus species have just two mating types, but others have thousands.

When two filaments meet, they begin an elaborate pas de deux, coordinating their dance with alternating chemical whispers. The opening sequence involves one filament's sending out a chemical that is unique to its own mating type. If its partner is of the same type, the dance ends and the filaments ignore each other. But if the partner is of a different mating type, the chemical binds to the filament's surface, causing it to respond by releasing its own chemical signal. Both filaments then sprout sticky outgrowths that grasp each other and draw the filaments together. The filament cells synchronize their cellular machinery and melt into each other to make a new individual.

The new fungus is an amalgam of its parents, but the fusion is not quite complete. The genetic material of the parents remains separate within the fungal body, existing as two distinct sets of DNA inside the cells. The mushroom maintains this united-but-separate arrangement throughout its feeding life belowground and even in the fruiting body that rises up to release the spores. Only in the gills that hang below the mushroom's cap does the full genetic fusion finally take place, after weeks or years of separation. But the union is brief. Immediately after the genetic material joins, it divides twice over to make spores that break loose and eject from their birthplace. Each spore will blow away in the wind or be carried by an animal to start the life cycle again.

The shaggy scarlet cup and other sac fungi follow a similar pattern, but their filaments don't join together until they are ready to make spores. The majority of their lives are spent belowground as unfused filaments. Only in adulthood do they seek another mating type with which to join, then grow a cup and produce spores.

The complexity of fungal sexual identity highlights the curious nature of the sexes in other kingdoms of life. Without exception, reproduction in animals and plants involves sex cells that come in two distinct forms: large and well-provisioned cells—eggs—or small and mobile cells, sperm. But fungi show us that this duality is not the only possible arrangement. Fungal mating types can number in the thousands.

The relative simplicity of fungus bodies may explain why they have not evolved specialized sperm and egg cells. The large, complex bodies of animals and plants take a long time to develop, and so they must start life with enough food supplies to complete their early development. But fungi have no elaborate bodies to build. Their simple filaments hatch fully formed from tiny spores. Producing an egg would be a waste of energy and time. The algae provide a good test case for this idea. They come in a wide variety of body forms: some are very simple, like fungi, whereas others have complex bodies, like plants or animals. As expected, simple algae have sex cells that are the same size, but complex algae have sex cells that have specialized into sperm and eggs.

Fungi may eschew the sexual roles of the rest of the multicelled world, but they still experience sexual divisions, with reproduction being possible only between individuals of different mating types. This seems wasteful. From the perspective of a fungal filament looking for a mate, the existence of mating types would seem to be a major hindrance, removing from the pool of potential mates up to half the other individuals in the species.

The puzzle of mating types has yet to be fully solved, but it appears that the politics of life within the cell may be at least part of the answer. Fungal cells are built on the same Russian Doll design as the cells of animals and plants. Fungi contain mitochondria that provide energy for the cell by burning food. In normal circumstances, the relationship between the mitochondria and their host cells is cooperative. But conflict waits just offstage.

Because mitochondria are the descendants of ancient bacteria, they retain their own DNA and multiply within the cell just like free-living bacteria. This multiplication is normally adjusted so that each cell has just the right number of mitochondria. But if things go wrong, an overgrowth of mitochondria will damage the cell. One way in which such unhealthy proliferation can happen is if mitochondria from two different fungi meet within a single cell. Under these conditions, competition among the different strains of mitochondria would favor those that divide most vigorously. Thus, a shortsighted struggle among mitochondria can destroy the longer-term success of the whole cell.

The mating types of fungi seem to be designed to prevent this kind of conflict. Mating types come with a set of rules specifying that only one mating type will provide mitochondria to the next generation.

Therefore, mating types provide a way for fungal cells to quash potentially damaging conflicts between mitochondria.

But theories about the origins and the evolution of mating types are uncertain and much debated. The fungi exhibit such a wide array of reproductive methods that most attempts at unifying explanations have foundered. For example, a few fungi produce structures that seem almost egglike, perhaps confounding the general rule that fungi don't produce eggs and sperm. In other species, mitochondria from different parent filaments sometimes mix together, breaking the rules about mating types. This diversity can be overwhelming, as students of fungal biology soon learn. But it also serves as a refreshing counterpoint to the rather uniform adherence to the roles of male and female in animals and plants.

From my prostrate position, I see hundreds of small cups and mushrooms spread across the surface of the mandala's leaf litter. Every decaying twig has one or more clusters of colored cups. Tiny brown mushrooms crown most of the dead leaves. That so many species and individuals could suddenly appear from a forest floor that I have gazed at for months is a reminder of how much of the forest's life is invisible to us, even with close observation. But unseen does not mean unimportant: these are the engines of decay, keeping nutrients and energy moving through the forest ecosystem. The lush summer productivity of this forest depends on the vitality of the underground fungal network.

July 13th—Fireflies

My body is tense as I pick through the misty air on my way to the mandala. I'm walking in the partial darkness of dusk. I place my feet carefully, straining my eyes through the gloom, searching for snakes in my path. Copperheads, *Agkistrodon contortrix*, the "hooked-tooth twister," concern me most. These snakes are particularly active on muggy summer evenings. Tonight, the copperheads' favorite summer snack is emerging. Hundreds of cicadas crawl up from their underground larval burrows. Snakes are surely on the prowl. I am loath to scorch my eyes with the reflected glare of a flashlight, so I move slowly, seeing the copperhead's leafy camouflage everywhere in the failing light.

My fear of predators was likely imprinted on my psyche by millions of years of natural selection. Tropical primates with poor night vision seldom live long if they have cocky attitudes toward the dark. Like all other living creatures, I am the descendant of survivors, so the fear in my head is the voice of my ancestors whispering their accumulated wisdom. My conscious mind chimes in with zoological fear-mongering: long-hinged fangs, painful blood-destroying venom, a pit near the eye that catches minute changes in temperature, a strike that lashes out in a tenth of a second. I reach the mandala, and its familiarity eases my tensions. Another whisper from the family tree: what is known is safe.

A firefly's flash greets me as I sit. The green light rises sharply several inches, then holds steady for a second or two. The evening has just enough daylight for me to see the beetle as well as its lantern. After the green glow dims, the animal hangs motionless for three seconds, then swoops down and across the mandala. The beetle then repeats the quick glowing ascent, lightless pause, and plunging departure.

If I were a real firefly connoisseur, I could identify the firefly's species by the distinctive rhythm and length of its flash, but such skills sadly elude me. During the day, I have used my field guide to identify fireflies from the genus *Photuris* clambering over the mandala's vegetation. Nightfall is too far advanced for me to pick out whether or not this individual is a *Photuris*, but the rising flash identifies him as a male. His flash is the opening line to what he hopes is a conversation with a future mate. He throws the line across the leaf litter, casting for the response that too often fails to come. After his flash, the male scans the forest floor, holding steady to give the female a chance to respond, then flies off to continue his search. Occasionally a female will return a flash from her hiding place, and the male will fly to her, flashing again. The pair signal back and forth several times, then mate.

If the firefly over the mandala is a *Photuris*, his mate will have an extra trick of the light to perform after they have bred. Once a female *Photuris* has finished the unexceptional task of luring suitors and mating, she turns her attention to the males of other firefly species. The unique flashing sequence of each species usually keeps males and females of different species apart. Just as we have no interest in the sexual signals of gorillas, fireflies ignore flashes from species other than their own. But *Photuris* females mimic the answering signals of other species, drawing in hopeful but hapless males, then seizing and devouring them. After walking down the aisle, the grooms become the wedding feast; the bride who seemed so appealing from afar turns out to be a very hungry gorilla. The femme fatale uses her prey not just for food but as a source of defensive chemicals. She steals these noxious molecules from her victim, then redeploys them inside her own body. If a spider catches her, she bleeds out the chemicals, repelling her attacker.

The forest floor, it seems, is full of hooked-tooth danger on these warm summer evenings.

But danger is only part of the story. Fireflies bring delight also, enchanting us with their sparkle and glow. Like the brilliance and hue of flowers, or the exuberance of birdsong, the twinkle of fireflies opens a window, blowing away the mist that stands between us and a truer experience of the world. When laughing children chase after fireflies, they are not pursuing beetles but catching wonder.

When wonder matures, it peels back experience to seek deeper layers of marvel below. This is science's highest purpose. And the firefly's story is rich in hidden wonder. The beetle's flash invites admiration for evolution's ability to cobble together a masterpiece from unremarkable raw materials: the lantern at the tip of the firefly's abdomen is made from standard-issue insect materials but assembled in such a way that the insect becomes a glowing forest sprite.

The insect's light flashes from a substance called luciferin. Like many other molecules, luciferin will combine with oxygen and turn into a ball of energy. The ball eases its excitement by releasing a packet of energy in motion, a photon that we perceive as light. Luciferin is similar in structure to other household molecules in the cell but, presumably through several mutations, has become particularly susceptible to overexcitement and relief. The molecule is assisted by two other chemicals whose job it is to whip luciferin into an overstimulated state.

Fireflies have therefore supercharged their internal chemistry to turn a glimmer into a glow. But chemicals alone produce at best a weak, diffuse light. The firefly's lantern is arranged to focus this potential into the flashes and counterflashes with which the courting fireflies so carefully time their prenuptial conversation. The lantern achieves this control by regulating the flow of oxygen to luciferin. Each cell in the lantern buries luciferin molecules in its core, then surrounds them with a thick mat of mitochondria. The usual function of mitochondria is to provide power for the cell, but the firefly's lantern uses them as oxygen sponges. Under normal conditions, any oxygen that seeps into these cells is quickly burned up in the mitochondria, leaving none to reach the core and stimulate luciferin. This layer of mitochondria is the firefly's "off" switch. When the time comes to flash, a nerve signal shoots into the lantern and causes a gas, nitric oxide, to flush out of the cells at the nerve tips. The gas shuts down the mitochondria, and oxygen blasts into the interior, igniting the chemical glow.

The firefly's flashing mechanism takes two ubiquitous features of animal physiology, mitochondria and nitric oxide, and combines them into an elegant and, as far as we know, unique light switch. The lantern's architecture is likewise a tinker's triumph, turning ordinary cells and the insect's breathing tubes into an airy home for luciferin. The tinker's work is no slipshod job. Over ninety-five percent of the energy that is used in the firefly's flash is released as light, a reversal of the performance of human-designed lightbulbs that waste most of their energy as heat.

In the sky above me night's darkness is complete. But as I stand to leave the mandala I see a forest full of lights. The fireflies stay within two or three feet of the ground, and from my standing position I look down on a swaying surface, a sea of glowing buoys. I light my path past imagined copperheads with my own lantern, pondering the contrast between my flashlight's inefficient industrial design and the biological wonders that dance all around me. But this is an unfair contest. I am comparing an infant with a sage. Our flashlights have barely two hundred years of thought behind them and have developed in a sea of abundant fossil and chemical energy. Humans have applied little effort to improving the prototypes of our first electric lights. With limitless fuel, why should we? In contrast, millions of years of trial and error stand behind the firefly's design. Energy has been in short supply for the beetles all along, producing a lamp that wastes little and uses beetle food, not mined chemicals, as its fuel.

July 27th—Sunfleck

It is midafternoon, but deep shade weighs on the mandala. The nadir of the year's cycle of daytime brightness has arrived. Now that summer is at its peak, the mandala's surface is darker than at any other time of the year. Even the winter solstice is brighter at ground level than is July's gloom. Greedy layers of maple, hickory, and oak leaves suck the sun's rays, stealing all but a fraction of a percent of the light that hits the canopy. Times are hard for the forest herbs; no wonder so many hurry through their yearly business in a few weeks of sunny springtime. Those low-growing plants that have not retreated into dormancy are adapted to lean living, scrounging light with leaves designed to persist on scraps. These forest herbs are the rangy desert goats of the plant world, with small appetites and thrifty flesh.

Suddenly, a column of intense light slants through the haze, beaming through a chink in the canopy and illuminating a single mayapple leaf in the mandala below. The mayapple shines in the spotlight for five minutes, then the beam's slow swing picks out a maple seedling, then another. Over the course of an hour the circle of brightness crawls over a *Hepatica*'s tri-lobed glossy leaf, onto sweet cicely, up into the spicebush, then across the jagged leaves of leafcup seedlings.

No plant gets more than ten minutes in the sun's eye before it is again covered by the blanket of shade. Yet fully half the plants' daily ration of light may arrive during the sunfleck's brief visit. The goats are given a few minutes at the feed trough before returning to the desert. But a bonanza of food can bloat and kill a hungry goat. Likewise, this sudden illumination is a mixed blessing for the mandala's plants. A dearth of light is a hardship that may eventually weaken a plant, but a sudden excess can wreck the leaf's thrifty economy, permanently impairing its function. Leaves in a sunfleck must therefore speedily rearrange their bodies to accommodate the sun's blast of energy.

Leaves are of course designed to snare light's energy and put it to work. They do this by deploying light-harvesting molecules that catch sunbeams and turn them into excited electrons. These electrons are whisked away, and their sparkle is used to power the plants' food-making machinery. But when too much light hits an unprepared leaf, energized electrons cannot be processed fast enough, and they wash around the delicate light-harvesting molecules, overwhelming them with their undirected agitation. Like a one-volt motor plugged into the wall, the leaf gets zapped. Plants adapted to shade are particularly vulnerable to damage from restless electrons. They have many more light-harvesting molecules than electron-processing molecules, so a sunfleck could easily overwhelm their inner architecture.

To cope with the arrival of a sunfleck, plants unplug some of their light-harvesting molecules before they can gather too much energy. At the first sign of trouble, an essential piece of the harvesting apparatus temporarily moves away from its usual location, returning only when the situation has calmed. This is like cutting a wire within an electric motor, stalling the motor's operation, then rejoining the wire's ends to start the motor again. A buildup of electrons also causes the stack of membranes that hold light harvesters to loosen, allowing energy to flow to the interior where electron processing takes place. The chloroplasts that contain all the photosynthetic machinery respond to the sunfleck by rolling to the edges of the cell, turning their faces away from the sun. In this way they protect the molecules within. When the sunfleck passes, the chloroplasts move back to the cell's upper surface, basking like lily pads in the forest's weak light.

The plants' response to the sudden influx of light is paradoxical. They unplug and roll away, seeming to shun the very thing that they have been seeking. The mandala's herbs sip at a mean trickle of light for most of the day, then hide their mouths under an umbrella when the deluge comes. But such is the force of the sunfleck's downpour that water splashes under the umbrella's rim, and plants receive a mouthful of life.

The sunfleck's sweep across the mandala illuminates everything in its path. A spiderweb glows silver in the glare, its invisibility ruined by the bright light. The leaf litter turns sandy bright and jumps into relief as dark shadows emerge. Iridescent wasps and flies shine like metal shavings scattered across the mandala.

The mandala's insects seem drawn to the circle of light, staying within its bounds as the sunfleck moves over the mandala. The most fastidiously loyal of these insect followers is a group of three ichneumon wasps. When a wasp steps out of the brightness, it immediately turns and scuttles back. The flies that also scuttle over the mandala have a looser attraction and make forays that last a minute or more into the darkness.

The sun-worshipping wasps overflow with nervous energy. They run frantically from side to side, constantly flicking their antennae and wings. They run their quivering antennae over and under every leaf in the sunfleck's small world. Every minute or two the wasps flip onto their sides and shudder their legs together, cleaning away the silk that spiders have strewn over the mandala. After the rubdown, the wasps jump back onto their feet and start over on their tremulous way.

The wasps' frenzy has a sharp purpose. They hunt for caterpillars on which to lay their eggs. Wasp larvae will creep out of the eggs and bore into the caterpillars' flesh, then larva will eat caterpillar, slowly, from the inside out, leaving the vital organs until last. The caterpillars live stoically on, feeding and digesting leaves even as their lives are stolen from within. These hollowed-out caterpillars therefore make excellent hosts, continually replenishing what the parasite robs.

The wasps' parasitic life cycle inspired one of Charles Darwin's more famous theological comments. He thought the ichneumon's trade was particularly cruel. These wasps seemed incompatible with the God he knew from his Victorian Anglican training at Cambridge. He wrote Asa Gray, the Presbyterian botanist at Harvard, "I cannot persuade myself that a beneficent and omnipotent God would have designedly created the Ichneumonidae with the express intention of their feeding within the living bodies of caterpillars." For Darwin, these wasps were the "problem of evil" writ in the script of the natural world. Gray was unconvinced by Darwin's theological arguments. While he continued to support Darwin's scientific ideas, he never abandoned his belief in the compatibility of evolution and traditional Christian theism. But suffering weighed heavily on Darwin; his body was always ill, and his spirit was bruised by the early death of his favorite daughter. As the dark years wore on, the weight of the world's pain pushed him from vague deism to skeptical agnosticism. Ichneumons were a symbol of the suffering he carried within, and their existence made a mockery of the God whose providence the Victorians saw written all over the natural world.

Theologians have tried to answer Darwin's challenge, but theistic philosophers have, perhaps unsurprisingly, little insight into the lives of caterpillars. Caterpillars are assumed to have no souls or consciousness, so their suffering cannot be a mechanism for their spiritual growth or a consequence of their free wills. Another argument claims that caterpillars don't really feel anything or, if they do, their lack of consciousness means that they cannot think about their pain, so the pain isn't true suffering.

These arguments miss the point. Indeed, they are not arguments but restatements of the assumptions that are being challenged. Darwin's claim is that all life is made from the same cloth, so we cannot dismiss the effects of jangling nerves in caterpillars by claiming that only *our* nerves cause real pain. If we accept the evolutionary continuity of life, we can no longer close the door to empathy with other animals. Our flesh is their flesh. Our nerves are built on the same plan as insect nerves. Descent from a common ancestor implies that caterpillar pain and human pain are similar, just as caterpillar nerves and human nerves are similar. Certainly, caterpillar pain may differ in texture or quantity from our own, just as caterpillar skin or eyes differ, but we have no reason to believe that the weight of suffering is any lighter for nonhuman animals.

The idea that consciousness is a humans-only gift likewise has no empirical basis: it is an assumption. But even if the assumption were correct, it would not resolve Darwin's ichneumon challenge. Is suffering greater when pain is embedded in a mind that can see beyond the present moment? Or, would it be worse to be locked in an unconscious world where pain is the only reality? A matter of taste, perhaps, but the latter option strikes me as the poorer one.

The sunfleck has swung across the mandala and now shines on my legs and feet. It moves on and beams directly on my head and shoulders, like a caricature of divine inspiration. The Sun Goddess unfortunately sends no sudden insight into the knots of philosophy; rather, she starts the sweat running down my face and neck. I'm feeling the energy that sustains the wasps' fidgeting dance across the forest floor. Their bodies are so slight that even a few seconds in the sun will raise their temperature by several degrees. To keep from roasting, the wasps send air currents flowing over their bodies, keeping a second-by-second balance between the inpouring of the sun's rays and the outflow of heat by convection. My own oozing sweat is the sluggardly response of a bulky mammal for whom heat balance is measured in hours, not seconds.

The sunfleck finally falls off my right shoulder, leaving the mandala as it travels east. The troubling wasps move with it. As the sunfleck flows away, dimness returns to the mandala, and I find that my senses have been changed by the experience of the sunfleck's passing. Now as I gaze around the forest I see not the uniformity I knew before but constellations moving over a dark sky.

August 1st—Eft and Coyote

Rain has drawn the leaf litter's humid world into the open. The litter's inhabitants scuttle exposed on the mandala's water-glazed leaves. The largest of these explorers is a salamander, a red eft, which stands on a mossy boulder, peering into the haze.

The eft's belly and tail rest on the rock. The animal's chest curves up, held by a push-up of spread front legs. The head is level and still. Eyes like droplets of gold stare unmoving across the mandala. Unlike the skin of most salamanders, the eft's looks dry, like crimson velvet, even in the heavy mist.

Two rows of bright orange spots run down the eft's back. These spots beam warnings to birds and other predators: stay away, toxins! The eft's skin is impregnated with poisons, giving the animal a shield against predation that most other salamanders lack. Efts are therefore confident, sauntering aboveground while most salamanders skulk below. This boldness explains the eft's unusually dry skin. Unlike their timid, light-fearing cousins, efts have thick, relatively waterproof skin that can withstand the daylight glare.

The eft holds still for a couple of minutes, breaks its trance with five steps across the moss, then halts and freezes again. Most likely it is searching for gnats, springtails, or other small invertebrates, using alternating bouts of quiet watching and surging movement to sneak up on, flush, and grab its prey. This is a common hunting tactic. Watch a robin on a lawn or a human searching for a lost cat and you'll see the same pattern of movement.

The eft's walking style is clumsy. Legs sprawl away from the body and oar the ground. A back leg swings out and forward, then the front leg on the opposite side, then the other back leg. The spine curves from side to side as the legs move, throwing the legs out and forward. This horizontal sway of the spine is like a fish swimming. Although the bones and muscles of efts are adapted to a terrestrial existence, their overall walking style is a fishy wiggle. This sideways twist works well for animals swimming against the all-encompassing solidity of water or soil, but on two-dimensional surfaces the writhe is inefficient—salamanders have to balance on three legs (or on their bellies) as they swing out one leg at a time. A panicked, running salamander is a whir of flailing limbs.

Terrestrial vertebrates whose lives require speed have reworked the fishes' ancient architecture at least three separate times. The ancestors of mammals and two lines of dinosaurs each came up with modifications to the sprawling inefficiency of the fish-on-land. Legs moved in and under, putting the animal's weight directly over its feet. This made it easier to balance and, therefore, to run without toppling over. The spine's side-sway was replaced with an up-and-down flex. Mammals are masters of this flex and can reach forward with both forelegs while pushing off with the combined power of both hind legs, then curve the spine down and stuff their forelegs back while swinging the hind legs forward to plant them ready for the next push-off. No salamander can match the bounding gait of a mouse, let alone the enormous leaps of a running cheetah. This newfangled spine has, ironically, returned to the ocean to compete with the old fishy spine. Whales move their tails up and down, rather than side to side, revealing their terrestrial ancestry. Mermaids, it seems, do the same.

The eft's spine and limbs make it ungainly on land, but the animal's life cycle is only partly terrestrial. "Efts" are just one of many stages in the life of the eastern red-spotted newt. The eft is a midlife stage, sandwiched between a larval stage and an adult. Unlike the eft, both the larva and the adult are aquatic. The larva chews its way out of an egg anchored to submerged vegetation in a pond or stream. The hatchling has feathery gills on its neck and lives submerged in water for several months, feeding on small insects and crustaceans. In late summer, hormones bewitch the larva's body. Gills dissolve, lungs grow, the tail turns from a paddle to a rod, and the skin roughens and flushes. The eft that walks onto land has been torn apart and rebuilt by an exaggerated puberty.

Once metamorphosed, efts stay ashore for one to three years, exploiting the bounty of the forest with no competition from bullying adults. Efts are like caterpillars, growing fat on a food source that no other life stage of their species can use. When efts get large enough, they return to water and transform themselves once more, this time into olive-skinned swimmers with sexual organs and keeled tails. These adults remain in the water for the rest of their lives, breeding yearly and, in some cases, surviving for more than a decade in this final life stage.

The complexity of this life cycle sheds some light onto the strange name of the animal in the mandala. *Eft* is the Old English name for newt, and this archaic label is retained to distinguish the immature terrestrial life stage from the sexually mature aquatic stage. Egg, larva, eft, adult; a succession that sends us to the basement of our language to rummage for words with which to tag all the phases.

When newts return to water to breed, their poisoned skin allows them to live side by side with large predatory fish, thus unlocking habitats that are too dangerous for other, less toxic salamanders. By damming streams and creating thousands of ponds stocked with bass and other predators, humans have unwittingly given the newt a great advantage over its salamander kin. The newt is riding the bow waves of the Great Ship of Progress.

The newts' repeated transformations are just one slice of the remarkably diverse repertoire of salamander life cycles. The *Plethodon* that squirmed across the mandala in February passed its larval stage in the egg. The egg hatched into a miniature adult that had no further metamorphosis to endure. So, *Plethodon* salamanders never need go to water to breed. Upstream from here, spotted salamanders lay eggs in ephemeral spring pools. Their larvae stay in the water, feeding desperately to transform into subterranean adults before their pond dries up. The streams that I can hear from the mandala contain two-lined salamanders that keep the egg-larva-adult system but remain in the stream as adults. Downstream from here mud puppies live in larger streams and rivers. They skip the "adult" stage, keeping the gilled larval body throughout their lives and growing reproductive organs in this juvenile form. Flexibility of sexuality and growth is therefore responsible for a large part of the salamanders' success. They mold their lives to fit the environment, and they live in a wider array of freshwater and terrestrial habitats than any other group of vertebrates.

As the standard-bearer of sexual flexibility lumbers out of sight, the mandala is washed with sounds from another champion of adaptability. A jumble of high-pitched barks and yowls is answered by a lower howl and yap. Then, the sounds meld into one tangled chorus of yowls and yips. Coyotes. They are very close. I am likely overhearing a mother greeting her teenage pups in the rock scree thirty paces east of the mandala.

The coyote pups were born in early April, just as the maples leafed out. Their parents courted and mated in midwinter and, unusually for a mammal, the male stayed with the female through her pregnancy and brought food for the youngsters for several months after they were born. By now the pups are old enough to have abandoned the cave, hollow log, or burrow that the mother chose as a nest site. Coyote parents leave the half-grown pups at rendezvous sites where the youngsters loiter and play while the parents search for food. Feeding adults often travel a mile or so away from the pups, then return amid joyful howls at dawn and dusk to feed, groom, and rest with their offspring. Most probably, it is one of these reunions that I heard. The weaned pups are first fed regurgitated food, then unchewed scraps. Over the late summer and fall the pups will range farther on their own, eventually leaving the natal home range in late fall or winter to seek their own home. Finding suitable territory that has not already been claimed can be hard, so these wanderers travel tens, sometimes hundreds of miles from their mothers' dens.

The air over the mandala has only recently been tickled by the yodels of coyotes. Although coyotelike animals may have lived here tens of thousands of years ago, these proto-coyotes were extinct long before the arrival of humans. When humans arrived, from Asia and, later, from Europe and Africa, coyotes lived in the western and midwestern prairies and scrublands; wolves ruled the eastern forest without interference from their smaller cousins. In the last two hundred years, however, wolves have declined rapidly and, in just the last few decades, the coyote has colonized the entire eastern half of the continent. What accounts for the startling reversal of fortunes of these two canids? Why did European colonization of North America crush the wolf but cause the coyote to sweep victorious over half a continent?

The wolf's symbolic role in European culture predestined North American wolves for intense persecution. The sound of howling wolves that woke the Pilgrims on the *Mayflower* during their first night in the "New World" stirred deep "Old World" fears. Wolves also lived in Europe, and their presence had soaked into the mythology of the colonists. Europeans regarded wolves as fearsome and turned the animal into a symbol of unleashed evil, nature's passions directed against us. As European wolves were exterminated, the distance between man and wolf widened, and the extravagance of the fear increased beyond anything justified by the wolves' depredations. When the *Mayflower* anchored at Cape Cod, the Pilgrims were therefore primed to shudder at the eerie howls. Here at last was the animal they had been taught to fear but had never encountered. At the time of the *Mayflower*'s journey, wolves had been extinct in England for more than a century, but in this savage new world they were, it seemed, everywhere.

This loathing is not entirely irrational. Wolves are carnivores whose specialty is eating large mammals. Because they hunt in cooperative packs they can easily take down animals heavier than themselves, including humans. We are their prey, so our fear is justified. Wolf behavior fans these flames of fear. Wolf packs trail lone human travelers for days, perhaps planning a kill, perhaps not. Such behavior guaranteed the wolf a position in our culture's symbology of evil. The fact that humans rank very low on the wolf's list of food preferences makes little difference. A few attacks and stalkings were enough to cement the Big Bad Wolf into our stories.

Direct persecution with traps, poisons, and guns accounts for a large part of the wolf's demise in North America. But Europeans also unwittingly launched an assault on the predator from another, indirect angle. Our rapacious use of wood and overharvesting of deer turned eastern North America from a meat-filled woodland into a deerless patchwork of farms, towns, and ragged logging scars. The champion predator of large herbivores was backed into a corner. The livestock that grazed in the formerly forested areas were the only prey left, and wolf attacks on homesteads increased the hatred, hardening the settlers' determination to stamp out the animal. Wolf eradication quickly became the policy of the new governments. States hired hunters, paid bounties, and, in a move that attacked both wolves and Native Americans, required "Indians" to pay, under penalty of "severe whipping," a yearly tax of wolf pelts. Wolves sat at the apex of the forest food web, a mighty but precarious position. Fated by their own specialization and by the fears of the colonists, they succumbed as the web was rewoven in the image of northern Europe.

Coyotes prefer to dance over the food web rather than perch atop it. The ax, plow, and chainsaw created forest clearings, pastures, and scrubby edges that offered the coyote just what it needed: plenty of rodents, berries, wild rabbits, and small domesticated animals. Coyotes are flexible, not fierce, and the loss of any one food item makes little difference to their ability to survive. They can hunt alone or in small groups, changing their social system to fit the environment. Wolf eradication removed another barrier. No longer would wolves persecute and hold back the lithe western invader.

Unlike top predators such as wolves, coyotes are abundant, and this makes them particularly invulnerable to attempts at eradication. As the French Revolution discovered, and the predator control arms of the U.S. federal and state governments rediscovered, it is harder to stamp out the upper classes than it is to kill the king.

The coyote also lacks the cultural baggage with which the wolf is saddled. No terrifying tales from Europe were pinned onto this North American native. Coyotes do prey on livestock, but they leave humans alone. So, although sheep farmers will kill coyotes and lobby the government to do the same, no coyote howl ever awakened the loathing of town dwellers, and no father ever hunted down a coyote for fear that it would slaughter his children as they played in the yard.

Coyotes swept into the northeastern corner of North America in the 1930s and 1940s. The southern wave started later, in the 1950s, and reached Florida by the 1980s. Coyotes arrived at the mandala sometime in the 1960s or 1970s, about a hundred years after the two native wolf species, red and gray, disappeared. Farther west, the invading coyotes overlapped with the declining wolves and may have picked up some genes from lonely remnants of the wolf population. Many of the first southern coyotes were surprisingly red and large, perhaps indicating a mixed parentage of coyote and red wolf. Analyses of DNA from living wolves and coyotes, and from museum skins that predate the coyotes' advance, support the idea that coyotes interbred with both gray and red wolves. The coyotes howling next to the mandala may, therefore, have a wisp of wolf in their blood.

Biological fluidity allowed the coyotes to flow into the hole left by the wolf. As deer became more abundant, coyotes spread out from the scrublands and into the forest. Eastern coyotes are larger than their western ancestors and, in some northern areas, have narrowed their diet and started specializing in deer. Coyotes have always preyed on fawns, but these new, larger coyotes hunt in packs and can bring down a healthy adult. It seems that the spirit of the wolf is returning, carried by the changing bodies of its coyote kin and perhaps helped by some stray wolf genes.

The coyote's colonization of the East has been a dance with the forest. The coyote's diet and behavior have turned and swayed, following the rhythm of the East. The dance partner, the forest, has added new steps and recovered some older, almost forgotten moves. Deer now have a wild predator, another layer of danger to add to disease, feral dogs, automobiles, and firearms. The catholic diet of coyotes means that their effect on the forest's choreography spreads beyond predation on deer. Fruiting plants now have an additional disperser, one that carries seeds many miles. Smaller mammals now live in fear of the wild canid. Coyotes also reduce populations of raccoons, opossums, and, to the consternation of pet owners, domestic cats. The suppression of these small omnivores has an unexpected silver lining for birds. Areas with coyotes are safer places for songbirds to build nests and raise young.

The addition of the coyote to the forest's troupe therefore sends ripples and lurches throughout. The predator makes life safer for the prey's prey. No doubt other parts of the forest also feel tugs and pulls. Because the coyote prances across the food web, eating fruits, killing the rodents that eat fruits, and eating the raccoons that eat fruits and rodents, the coyotes' ecological effects are hard to predict. Is seed dispersal helped or hindered? How do ticks fare with fewer mice but more birds? The forest's future depends, in part, on the answers to these questions.

Coyotes also teach us something about the forest's past. The original dancers, the wolves, are gone, but their understudies, the coyotes, give us a glimpse at the former grace and complexity of the forest's motion. The deer, also, are filling in. They not only play their own role but have taken on the parts of the elk, the tapir, the woodland bison, and other extinct herbivores. The success of the coyote and the deer in the eastern United States is therefore both a symptom of our culture's profound effects on the forest and a return to a semblance of the cast and plot of the continent before the arrival of Pilgrims, guns, and chainsaws.

Although the mandala sits in an old-growth forest, the flow of life here is powerfully affected by currents running in from the surrounding landscape. The coyote owes its presence in the mandala to the cascade of changes that European colonization brought to North America. This cascade also affected aquatic ecosystems, and there would be fewer efts in the mandala if humans had not dammed nearly every stream around it, creating scores of ponds and lakes.

Ecological mandalas do not sit isolated in tidy meditation halls, their shapes carefully designed and circumscribed. Rather, the many-hued sands of this mandala bleed into and out of the shifting rivers of color that wash all around.

August 8th—Earthstar

Summer's heat has coaxed another flush of fungi from the mandala's core. Orange confetti covers twigs and litter. Striated bracket fungi jut from downed branches. A jellylike orange waxy cap and three types of brown gilled mushroom poke from crevices in the leaf litter. The most arresting member of this death bouquet is the earthstar lodged between rafts of leaves. Its leathery outer coat has peeled back in six segments, each segment folded out like a flower's petal. At the center of this brown star sits a partly deflated ball with a black orifice at its peak.

My eyes meander over the mandala's surface, delighting in the profusion of fungal bodies. Two white domes at the mandala's edge eventually catch my attention. The spheres emerge from the receding tide of decomposing litter. I reposition myself to get a closer look. Golf balls! Like a discarded beer can in a stream or bubble gum stuck into the bark of a tree, these plastic globes seem profoundly ugly and out of place.

The golf balls were sent flying from the high bluffs that overlook the mandala. A golfing friend tells me that shooting a ball from the edge of the cliff gives him a thrilling sense of power. The golf course extends to the cliff edge, providing ample opportunity to indulge this buzz. Most of the balls land to the west of the mandala where local children gather bagsful to sell back to the golfers.

Glossy white plastic balls are visually startling in the context of a forest. But the balls also jar because they have arrived from a parallel reality. The mandala's community emerges from the give-and-take of thousands of species; a golf course's ecological community is a monoculture of alien grass that emerged from the mind of just one species. The mandala's visual field is dominated by sex and death: dead leaves, pollen, birdsong. The golf course has been sanitized by the puritan life-police. The golf green is fed and trimmed to keep it in perpetual childhood: no dead stems, no flowers or seed heads. Sex and death are erased. A strange country.

A dilemma: should I remove the balls or leave them nestled in place? Removing them would break my rule about not meddling in the mandala. But taking them away would restore the mandala to a more natural state and might make room for another wildflower or fern. Discarded golf balls have nothing to contribute to the mandala. They don't decompose and release their nutrients. They don't become another species' habitat. The grand cycle of energy and matter seems to halt when it reaches a dumped golf ball.

My first impulse, therefore, is to restore the mandala to "purity" by removing the plastic balls. But this impulse is problematic for two reasons. First, removing the balls will not cleanse the mandala of industrial detritus. Acidity, sulfur, mercury, and organic pollutants rain in continually. Every creature in the mandala carries in its body a sprinkling of alien molecular golf balls. My own presence here has undoubtedly added strands of worn clothing fiber, alien bacteria, and exhaled foreign molecules. Even the genetic code of the mandala's inhabitants is stamped by industry. Flying insects, in particular those whose ancestors have come near humans, carry resistance genes for many pesticides. Removing the golf balls would merely tidy up the most visually obvious of these human artifacts, preserving an illusion of the forest's "pristine" separation from humanity.

The impulse to purify might fail on a second, deeper level. Human artifacts are not stains imposed on nature. Such a view drives a wedge between humanity and the rest of the community of life. A golf ball is the manifestation of the mind of a clever, playful African primate. This primate loves to invent games to test its physical and mental skill. Generally, these games are played on carefully reconstructed replicas of the savanna from which the ape came and for which its subconscious still hankers. The clever primate belongs in this world. Maybe the primate's productions do also.

As these able apes get better at controlling their world, they produce some unintended side effects, including strange new chemicals, some of which are poisonous to the rest of life. Most apes have little idea of these ill effects. However, the better-informed ones don't like to be reminded of their species' impact on the rest of world, especially in places that don't yet seem to be overly damaged. I am such an ape. Therefore, when a golf ball in the woods strikes my eyes, my mind condemns the ball, the golf course, the golfers, and the culture that spawned them all.

But, to love nature and to hate humanity is illogical. Humanity is part of the whole. To truly love the world is also to love human ingenuity and playfulness. Nature does not need to be cleansed of human artifacts to be beautiful or coherent. Yes, we should be less greedy, untidy, wasteful, and shortsighted. But let us not turn responsibility into self-hatred. Our biggest failing is, after all, lack of compassion for the world. Including ourselves.

Therefore, I resolve to leave the golf balls in the mandala. I'll continue removing strange plastic objects from the rest of the forest, but not from here. There is value in keeping a patina of "naturalness" along hiking trails and in gardens. Our harried eyes need a visual break from the productions of industry. Keeping the woods trash-free is a symbol of our desire to be more careful members of life's community. But there is also value in the discipline of participating in a world as it is, discarded golf balls and all.

Yet the utter indigestibility of the golf balls seems an affront to the mandala's other creatures. Eighteenth- and nineteenth-century golf balls were biodegradable, being made from wood, leather, feathers, and tree resin. Modern "ionically strengthened thermoplastic" balls cannot be eaten by bacteria or fungi. One billion golf balls are manufactured each year. Are they all destined for a brief bounce on the green, then eternal life as garbage? Not quite, is my guess. The golf balls in the mandala will continue to sink through the litter as the biological material they rest upon decays. In a few years they will hit sandstone and lodge between the jumbled boulders that underlie the mandala. Here they will be ground to ionically strengthened thermoplastic dust. The escarpment on which we sit is receding eastward, so the golf balls will join the slow rumble of grinding rock, and the little balls will be pulverized. Eventually their atoms will settle into new rock, either in a compacted layer of sediment or in a hot pool of magma. Golf balls don't end the cycle of matter, as they seem to do. They take mined oil and minerals into a new form, soar briefly, then return the atoms to their slow geological dance.

Another fate is possible. The earthstars and mushrooms that ring the mandala's golf balls may devise a way to digest and recycle the balls' plastic. Fungi are masters of decomposition, so natural selection might produce a plastic-munching mushroom. Stupendous quantities of matter and energy are locked up in plastic. Evolutionary triumph awaits the mutant fungus whose digestive juices can free these frozen assets and conjure them to life. Fungi, and their equally versatile partners in the business of rot, bacteria, have already shown themselves capable of thriving on other industrial innovations such as refined oil and factory effluent. Golf balls may be the next breakthrough. "Are you listening? Plastics. There is a great future in plastics."

August 26th—Katydid

CHA CHA! CHA CHA! The whole forest vibrates.

It is evening and the mandala is dim, unfocused, made of patches of light and dark. As light fades, the chorus pounds louder. *CHA CHA! CHA CHA!*—the double beat of thousands of katydids singing from the trees. Occasionally the isolated notes of a single singer stand out, but mostly individual triplets and couplets merge with the songs of others: *CHA!* The insects question the forest, then answer, "ka-ty-did? she didn't!," pause, then question and answer again. The exclamations tumble into one another, melding into a thumping beat. The rhythm holds steady for a minute or more, breaks into a din of unsynchronized songs, then unison is reestablished.

The barrage of sound is the acoustic expression of the forest's great productivity. Sun energy, turned to tree energy, turned to katydid energy. Katydid youngsters feed on leaves through the summer, gradually molting into larger sizes, finally emerging as thumb-sized adults. The great vigor of the forest's plants thus translates into spectacular blasts of sound. The katydid's scientific name expresses this connection, *Pterophylla camellifolia*, the camellia leaf-wing. Not only is the katydid's life powered by and built from foliage, but the insect looks just like a leaf.

Katydids sing with their wings. A corrugated ridge, called a file, runs across the base of the left wing, just behind the head. A nub on the right wing sits opposite the file. The insect strums these wing bases together, drawing the nub like a plectrum over the file to make a buzz or hum. Katydids are no amateur jug band strummers. They inflect the vigor, angle, and length of their strokes like master violinists on their bows. The katydids' speed outshines concert-hall virtuosos and backwoods flat-picking guitar champions. Some species strum more than a hundred times each second, which, when combined with the closely spaced bumps on their files, produces fifty thousand pulses of sound per second, sounds that are well above the limits of human hearing. The katydids around the mandala are more mellow strummers, pulsing just five to ten thousand sound waves each second. These notes are higher than the highest notes on a piano keyboard, but they are low enough that our ears can perceive their whines.

The katydid's file and plectrum do not work alone. The secret of the katydids' loudness is a patch on the wing that acts like the skin of a banjo, resonating and amplifying the plectrum's vibrations. This skin is tightened such that its resonant tone is different from the note produced by the file. These mismatched tunings produce a clash of vibrations that combine to make the katydid's dissonant buzz. Crickets, unlike their katydid cousins, have skins tuned perfectly to their files, allowing them to sing sweet notes unsullied by harsh side tones.

Like those of humans and many birds, katydid songs come in regional dialects. Northern and midwestern katydids sing slowly and with two or three syllables. *Ka-ty, Ka-ty-did, she did-n't.* Southerners add more syllables and sing much faster. *Ka-ty-did-n't, she-did-n't, did-she, ka-ty-did.* In the West, katydids sing slowly and with just one or two syllables. *Ka-ty, did, did, ka-ty.* Katy's story evidently has many interpretations. The function or consequences of these regional accents is unknown. Perhaps dialects adapt the songs to the acoustic properties of different forests? Or maybe they reflect hidden differences in the preferences of females in different regions, differences that may inhibit crossbreeding between populations with different ecological adaptations?

The katydid chorus is punctuated by brief, dying bursts from cicadas. Cicadas are singers of blazing hot afternoons, and they relinquish their acoustic dominance as dusk advances. The cicadas' drawn-out whine comes from an apparatus that is stranger yet than the katydid's plectrum, ridge, and drum. On each side of the cicada's body sit discs embedded in the hard external skeleton. The discs look like closely barred porthole windows. The bars are stiff struts that can snap back and forth sideways. When a muscle pulls on the disc, the struts snap in a cascade, producing a trill, then each strut pops back as the muscle is relaxed. The sound of every snap and pop is amplified by membranes and an air-filled sac inside the cicada's body. These corrugated discs, called tymbals, are unique in the animal world.

Both cicadas and katydids derive their energy from plants. Cicada larvae are subterranean tree parasites, sucking tree sap from roots, living like moles with syringes. Unlike the rapidly growing katydids, young cicadas take several years to mature. Tonight's cicada chorus is therefore fueled by four or more years of tree juice, sung by moles that have crawled from their burrows and taken to the trees.

Female katydids and cicadas wander the treetops, contributing no sound of their own but listening to the chorus of males. Katydids hear with nerves on their legs; cicadas have ears lodged in their abdomens. If the male singers are loud or snappy enough to lead the chorus, the listener will move closer, listen some more, then mate.

When female and male katydids entwine, he passes to her not only a small sac of sperm but a large "nuptial gift" of food. This food sac is usually about one-fifth the male's body weight. The manufacture of the sac is so taxing that the male's abdomen is mostly filled with food sac glands. The function of the gift varies from species to species. In some katydids the male is providing food with which the female will produce eggs. In other species, the gift prolongs the female's life span.

Unfortunately for singing male katydids, potential mates are not the only listeners in the forest. Singing undoubtedly increases the risk of being discovered by a bird. Cuckoos, in particular, are fond of hunting katydids. But the most abundant and dangerous enemy of songster katydids are tachinid flies. These spiny beasts feed on nectar as adults, but their larvae are parasites of other insects. Several tachinid species specialize in katydids and have ears tuned specifically to the song of their preferred host. The eavesdropping mother fly homes in on her prey, lands close by, then deposits a brood of writhing larvae. The larvae swarm onto the victim and burrow into his body. Like ichneumon wasps inside a caterpillar, the fly larvae slowly consume the katydid from the inside out. The hit-and-run strategy of mother flies is guided solely by sound, so tachinid parasitism is a burden carried almost exclusively by male katydids.

Darkness advances. The cicadas finally fall silent, ceding the chorus until tomorrow, when the day's heat will wake them. Other katydid species join in. Lesser angle-wing katydids shake out bursts of raspy sound, like arboreal maracas. Whines and buzzes from other species poke out from the chorus, hinting at the diversity of leaf chewers above.

As dusk progresses my sense of sight fails, and the forest swells around me in dark billows that finally merge into blackness.

Only the forest's exultant thunder remains: *CHA CHA! CHA CHA!*

September 21st—Medicine

I feel a powerful sense of delight in the morning's strong sun. My spirits were lifted by the sight of a dozen migrant warblers bathing in the stream that crosses my path to the mandala. They stood in shallow stream pools, dipping and shaking their bodies, feathers fluffed out. Each bird raised a halo of silvery flashing water drops. They seemed to baptize themselves with sunlight.

The birds' unrestrained enjoyment gave me particular gratification, because this same stream was the source of much recent trouble. Two days ago, on my walk from the mandala, I found the stream ransacked, every rock turned over or thrown aside. This had happened before, poachers coming through and grabbing every salamander they could, hauling them away for use as bait. The stream was gutted. The forest's salamanders would die on hooks or in stinking bait buckets. I felt disgust and visceral anger. I walked on, my ire surging and coiling into itself. Up the hillside I marched, mind wound tight. When I reached the base of the cliff, the tension snapped and unwound: my heart kicked and started fibrillating, the beats coming in uncountable squalls.

There followed a difficult bike ride into town, a few hours in the hospital with IVs and drugs. Within a couple of hours, my heart settled back down and, after a day of rest, I've returned to the forest. Today, the warblers' glistening beauty therefore seems particularly sweet, perhaps even redeeming.

At the mandala, I see the plants with new eyes. In addition to the ecological community, I now see a pharmacopoeia. This new way of seeing came to me with the drugs I was given at the hospital, both of which were derived from plants. Aspirin, originally from willow bark and meadowsweet leaves, slipped into my cells and, like the chemicals in mosquito and tick bites, disabled the processes that cause clotting of blood. Digitalis, from foxglove leaves, bound to my heart's cells, shifting the chemical balance, making my heartbeat stronger, surer.

In the hospital room, I first felt separated from nature, but this was an illusion. Nature's tendrils penetrated the room, reaching out to me through pills. Plants twined inside me, their molecules finding and grasping mine in a close embrace. Now I see these connections at the mandala: every species brims with medicinal potential. Willow, meadowsweet, and foxglove do not grow here, but the mandala's plants have their own healing properties.

Mayapple is one of the more common plants on this mountainside, and its umbrella-like leaves poke up from several places in the mandala. The ankle-high leaves of mayapple grow from underground stems on the forest floor. The stems grow horizontally, branching through the leaf litter, gradually expanding until dozens of leaves grow in a patch that can be several meters across. Native Americans have long known that the plant has powerful properties. At very low doses, extracts from the plant were used as laxatives and to kill intestinal worms; higher doses, which would be fatal if ingested by people, were put onto newly sown corn to protect the seed from crows and insects.

Modern studies of mayapple have found that the plants' chemicals can kill viruses and cancer cells. Mayapple extract is now used in creams that heal warts caused by viruses and, after the extract has been chemically modified in the lab, as chemotherapy against cancer. These drugs could obviously not exist without mayapple. Less obvious is their dependence on other members of the forest community. Bumblebees pollinate the flowers of mayapples, flying under the leaves to reach the nodding white blooms. Later in the summer, the flowers mature into small yellow fruits, each about the size of a small lemon, the "apples" for which the plant is named. Box turtles have an unusual affinity for these mature fruits, sniffing them out, devouring them, and then wandering away with a bellyful of mayapple seeds. Without passage through the gut of a turtle, the seeds generally cannot germinate. Pharmaceutical textbooks do not discuss the ecology of forest-dwelling bumblebees and turtles, but the practice of medicine needs these species nonetheless.

Wild yam is another local species with important medicinal properties. It is not found within the mandala's circle but grows scattered all around, particularly in wetter, shady patches of forest. Yam grows as a vine, wrapping its thin stem around shrubs and small trees as it climbs to head height or above. The stem and heart-shaped leaves are delicate and do not survive hard freezes, so the plant overwinters as fingerlike tubers under the leaf litter. These tubers are rich in chemicals that are structurally similar to human hormones, including progesterone. This fact was not lost on Native Americans, who used the plant to relieve the pains of childbirth. Later, in the 1960s, the first birth control pills were made by chemically modifying extracts from the tubers. Yams are also reported to lower cholesterol, reduce osteoporosis, and relieve asthma, although the evidence for these properties is debated.

Mayapple and yam are easy to find in this forest. Another wild medicinal species, ginseng, is unfortunately not so common. Its fate offers a caution about the overharvesting of useful wild plants. The human appetite for ginseng's stimulative and healing properties is so strong that most of eastern North America has been stripped of this once abundant forest herb. In the mid-nineteenth century, the United States annually exported between one-half and three-quarters of a million pounds of ginseng. Similar quantities may have been used domestically. Now, the annual export is less than a tenth of what it was, driven down by the rarity of the plant. Despite regulation of the ginseng harvest by the federal and state governments, the market for "sang," as it is known locally, is thriving. A few miles down the road from the mandala, dealers set up seasonal stalls at major road intersections and buy roots from local "diggers." Dried roots fetch five hundred dollars or more per pound, a price that provides strong motivation to search for new plants. For skilled diggers, the harvest provides a significant economic opportunity in an otherwise difficult local economy.

The decline in ginseng's abundance has induced some forward-looking dealers and diggers to start cultivating semiwild populations of ginseng by sowing seed in the forest as they hunt for roots. Like box turtles transporting mayapple seeds, humans have now taken on the role of seed dispersers. This task was previously performed by birds, especially thrushes, that regard ginseng's ruby berries as a tasty late summer snack. Fortunately for the human sowers, ginseng seeds are less fussy than those of mayapple and will germinate without passage through a bird gut. Whether these sowing efforts can protect ginseng from further declines is presently unknown; most botanists remain very concerned for the species' future.

Ginseng, yam, and mayapple are all small plants that overwinter as nutritious underground stems or roots. This shared way of living explains why they are all so rich in medicinal chemicals. Unlike fast-moving animals or thick-barked trees, these stationary, thin-skinned plants are highly vulnerable to attack from mammals and insects. Their underground stores of food are particularly attractive to would-be predators. Because they lack the ability to flee or to hide behind strong walls, the plants' only defensive option is to soak their bodies in chemicals that play havoc with the guts, nerves, and hormones of their enemies. Because natural selection has designed the defensive chemicals specifically to attack the physiology of animals, these poisons can, in careful human hands, be turned into medicines. By finding just the right dose, herbalists can turn the plants' defensive arsenal into an impressive collection of stimulants, purgatives, blood thinners, hormones, and other medicines.

The mandala's medicinal plants and the drugs in my bloodstream are representatives of a much larger group: one-quarter of all prescriptions are filled with medicines derived directly from plants, fungi, and other living organisms. Many of the remaining prescriptions involve modifications of chemicals originally found in wild species. But the complex chemical world of the mandala's species is poorly understood. Of the thousands of molecules in the two dozen plant species visible in the mandala, only a handful have been thoroughly investigated in the laboratory. Others have yet to be examined, despite their use in traditional herbal medicine. The invisible biochemical diversity of the mandala is full of potential that awaits exploration.

My experience with botanical medicines has taught me that my kinship with the mandala's inhabitants extends all the way to the tiny scale of molecules. Previously, my kinship primarily meant shared genealogy on an evolutionary tree and interconnected ecological relationships. Now I understand how intimately my physical being is tied into the community of life. Through the ancient biochemical struggle between plants and animals, I am bound to the forest through the architecture of my molecules.

September 23rd—Caterpillar

Migrating warblers eddy through the mandala's trees, washing through branches in waves. A Tennessee warbler, returning from its breeding grounds in the northern forest, drops to a low maple sapling at the mandala's edge and probes the leaves in search of food. The bird has another two thousand miles to fly before it reaches its winter home in southern Central America. Feeding is an urgent concern.

The state of the leaves above the mandala hints at the source of the warbler's meals. Each leaf has been shotgunned, leaving ten or more ragged holes. Most leaves have lost nearly half their surface area. The mandala's caterpillars have turned summer leaves into insect flesh. This flesh will, in turn, fuel the warblers' long journey.

Caterpillars are famously gluttonous. During their lives they increase their weight two or three thousandfold. If a human baby did the same, she would weigh nine tons, the combined weight of several marching bands, by adulthood. And, if the baby kept pace with the caterpillar, adulthood would arrive just a few weeks after birth.

Caterpillars grow fast because their being is focused on just one task: eating leaves. Unlike adult insects they do not grow tough external skeletons, wings, complex legs, sex organs, or elaborate nervous systems. Such accoutrements would blunt the caterpillars' focus and slow their growth. Defensive bristles are the only nongourmand elaboration that natural selection has allowed. By specializing on the task of feasting, caterpillars have opened a business in which they have few rivals. In most forests they consume more leaves than all other herbivores combined.

A fat tussock moth caterpillar shuffles into the mandala. The caterpillar is a carnival of colored plumes and hairs, its brightness advertising nastiness from stinging hairs and internal poisons. Four yellow tufts project from its back, like shaving brushes pointed at the sky. These sit in a haze of long silver hairs that sprout from each body segment. Two black sprays of hair jut from either side of the head, and the tail is tipped with a cluster of brown needles. Where it is visible through the haze of hairs, the caterpillar's skin is striped yellow, black, and gray: a gorgeous and fearsome outfit.

Adult tussock moths do not expose themselves to danger by munching leaves in the open. Thus, their colors can be bland. The female emerges from her hidden cocoon, then clings to it and waits for a male. She is flightless and looks like a furry sleeping bag. Because she has no need to wander, she need not advertise her distastefulness and can rely on camouflage to protect her. The adult male is a strong flyer. He sniffs out female pheromones with his feathery antennae, mating with her, then flying away. Both sexes are inconspicuously colored brown and gray, protected by the female's utter immobility and the male's vigorous wings. As is true for many other moths, natural selection's paintbrush has produced an audacious youngster and a dour adult.

As I watch the gaudy caterpillar, a black ant climbs onto its back, squeezing through the bristles like a man pushing through a thicket of bamboo. The ant pushes down its mandibles, straining unsuccessfully toward the caterpillar's neck. The caterpillar marches on, seemingly unperturbed by its attacker. The ant moves back from the neck and bites down between the yellow tufts, but again it cannot reach the skin. Then another ant, smaller and honey-colored, climbs aboard and joins the attack. The ants meet and fight each other, grappling on a mat of yellow hairs. The honey-colored ant is thrown off, climbs back on, falls off, and is followed by the black ant. The caterpillar accelerates, perhaps trying to escape, but the ants circle. The black ant lunges into the caterpillar and attacks again, biting down repeatedly but never making contact with the soft caterpillar skin. The ant falls off, and the caterpillar immediately climbs a dead leaf that arches above the forest floor. The caterpillar stops. Is it outmaneuvering the ants? The ants circle on the floor but cannot find the victim. Their circles finally carry them away from the leaf. The caterpillar climbs down and lumbers toward the trunk of a large maple just outside the mandala. Free!

A smaller tussock moth caterpillar was not so lucky. Ants drag its corpse away to feed their nest mates. Perhaps this caterpillar's hairs were too short or its evasive maneuvers too slow? Whatever the cause of its demise, it now joins a quiet funeral procession of dead caterpillars flowing into the maws of ant colonies in and around the mandala. One study counted more than twenty thousand caterpillars entering an ant nest each day. Until witnessing the caterpillar's struggle in the mandala I assumed that birds were the cause of caterpillars' hairiness. But hairs evidently also keep ant mandibles away from caterpillar skin. The scientific literature confirms what I observed today: ants are major enemies of most caterpillars.

One group of butterflies has turned this ant-agonistic relationship around. The blues, or lycaenids, have evolved a mutualistic relationship with ants. The caterpillars of blues are hairless and supremely vulnerable to ant attacks. But ants generally don't bite them, preferring instead to feed on the sweet "honeydew" that the caterpillars secrete for them. The caterpillars' gift to the ants is perhaps akin to mafia-style protection money. For the price of some sugar, the caterpillars are unharmed by the ants. But the ants don't simply withhold attack in return for food: they actively protect the caterpillars, fending off other predators, especially wasps. So a closer analogy might be that the ants are hired as bodyguards. Ants give lycaenid caterpillars a tenfold survival advantage relative to caterpillars without ant attendants. The caterpillars seem to prefer life with ants, and some have special scrapers that they use to produce vibrations on leaves. These vibrations attract ants, so the caterpillars literally sing for their guards.

Having escaped the ants, the mandala's tussock moth caterpillar ascends a maple trunk. There are no ants on the tree, but spiders have painted most of the trunk with sticky strands, and the caterpillar has difficulty pushing through. Patches of moss, still wet from last night's rain, present another challenge. The hooklets on the caterpillar's legs lose their grip, and the animal falls back several inches before struggling on.

The caterpillar's ascent carries it into a world dominated not by ants but by birds. Ants find their prey through touch and smell. Birds use vision. Pigments and patterns are therefore vitally important if one is to escape the attentions of birds. Because humans are so visual, we are fascinated by the extraordinary diversity of caterpillar art. Caterpillars feature prominently in children's stories, and many naturalists owe their love of nature in part to the caterpillars' charm. In contrast, the larvae of flies, wasps, and beetles live hidden from the keen eyes of our avian cousins, so these pasty white larvae have little allure for us.

The tussock moth caterpillar in the mandala uses arresting contrast between bright yellow and black to advertise its distastefulness. The brushes of yellow hair present a striking textured contrast to the spiny silver hairs that fuzz over the rest of the body. The display leaves the observer in no doubt that spines, hairs, and toxins are abundantly provided on this animal. Most birds will not even peck at such a display. Similar costumes are found in other poisonous or bristly caterpillars, each species creating its own variations on the themes of hue and contrast.

Caterpillars that lack spines or noxious chemical defenses are deceivers, not advertisers. They mimic bird droppings, dead leaves, twigs, small snakes, or poisonous salamanders. Natural selection used a refined hand as it turned out these animals, giving leaf buds to twig mimics, endowing snake mimics with eyes that have fake reflections in their pupils, and adding small droppings to the surface of leaf mimics.

The gaze of birds has not wavered from caterpillars for millions of years, turning caterpillar bodies into masterworks of visual design. Remarkably, this gaze has molded yet more. Even the pattern of light through the mandala's nibbled leaves is shaped by the birds' discriminating eyes. Feeding birds learn to associate ragged holes in leaves with the presence of caterpillars. Because leaves remain damaged long after caterpillars have moved on, birds continually update their feeding patterns based on their recent experience of feeding in particular tree species. Caterpillars that excise obvious holes in leaves, then linger next to these holes, will quickly attract the attention of these smart birds. Therefore, only well-defended caterpillars can afford to be messy eaters. Caterpillars that are more vulnerable to birds, such as those with few hairs, fastidiously pare leaves down from the edges, leaving no telltale holes, maintaining the silhouette of an entire leaf. Some caterpillars curl their leaf-mimicking bodies onto the missing leaf edges, filling out the leaf's profile and fooling predatory eyes. The leaves above the mandala bear the jagged marks of insouciance, and I suspect the tussock moth caterpillar and its kin were responsible for most of the damage.

Bird eyes have carved and painted the mandala. The form of both nibbling caterpillar and nibbled leaf reflects the evolutionary struggle between caterpillars and birds. The migrating warblers seem ephemeral, but their presence will outlast their physical departure.

September 23rd—Vulture

My study of the chewed leaves in the canopy has drawn my eyes skyward. The summer canopy usually narrows my world, directing my gaze down, but now I peer through chinks in the tree cover. The sky has been cleared of dust by a vigorous rainstorm that gusted through yesterday, leaving glassy blue. Summer's humidity has dropped, giving the day's heat a comfortable feel. This is September's typical weather, long stretches of open skies interrupted by warm, boisterous fronts, often the remnants of tropical storms from the Gulf.

Today, a turkey vulture circles directly above the mandala, its broad wings held like unflapping sails against the sky. After its pivot, the vulture soars eastward, puffed away by a sudden draft.

The mandala sits far enough south that turkey vultures are found here in every month. At this time of year, our local resident birds mix with northerners who migrate over Tennessee to winter on the Gulf Coast and Florida. Some birds winter even farther south, trekking to Mexico and beyond. These long-distance migrants will have company—turkey vultures are permanent residents of Central and South America, making them one of the most widely distributed species of birds in the New World.

Unlike most flying birds, turkey vultures are easy to identify, even from a great distance. They hold their wings in a shallow V with wingtips splayed up, like an airborne curly bracket }. They fly with a drunkard's gait, swaying and tipping as they go. This apparent lack of sobriety has an aerodynamic reason; turkey vultures are masters of soaring flight, seldom flapping their wings and almost never flapping more than ten times in a row. To ride the wind with such energy-saving ease, the great paddlelike wings catch updrafts and eddies, turning to the birds' advantage every rising puff of air. This results in a slow, teetering flight style whose superficial inelegance hides its extraordinary efficiency. The drunkard is a thrifty genius with no need for maneuverability, grace, or speed. Turkey vultures spend their days leisurely scanning their domain, riding the skies for up to a third of their waking hours.

The turkey vultures eat nothing but carrion, and their frugal flying style allows them to patrol tens of thousands of acres each day in search of corpses. In forested areas, where they do much of their feeding, the tree canopy obscures the view, but even with a clear line of sight, immobile bodies covered by camouflaged fur are hard to spot. Yet turkey vultures sleuth them out with thoroughness and precision. Dead chickens and rats deliberately placed in forests by scientists are usually found within a day or two, even when the bait is covered with leaves and brush. Turkey vultures are evidently sniffing their meals, using their wide nostrils to peer through the forest's confusion of color.

Finding a rank corpse by smell is hardly an impressive feat, but vultures do much better than this. Indeed, overly putrid meat is not much to their taste. Turkey vultures instead cruise the skies in search of the subtle whiff of recent demise. Unlike the barrage of odor from a festering carcass, the smell of fresh carrion is slight, made up of a few choice molecules exhaled by microbes and the cooling corpse. Soaring vultures catch these wafts and follow them to the ground, pinpointing their destination from the thousands of acres in their field of view.

In the modern world, the turkey vultures' sense of smell can sometimes lead them up dead ends (figuratively, for a change). They circle over abattoirs that, although they look like ordinary warehouses, send skyward the aroma of the recently departed. Pipelines offer similar frustrations. Gas companies add a small amount of an odorous chemical called ethyl mercaptan to the otherwise odorless natural gas in their transmission pipes. If a valve or pipe seam fails, the smelly chemical leaks out with the natural gas, alerting human noses to an explosion hazard. But vultures also smell the leak and will congregate around cracked pipes, becoming unwitting assistants in the search for pipeline flaws. This tangle of vulture and human noses is caused by the bouquet of death—ethyl mercaptan is given off naturally by corpses. Because humans have a deep aversion to rotting meat, our noses are extremely sensitive to ethyl mercaptan. We can pick out its smell at concentrations two hundred times lower than our threshold for ammonia, which is itself strongly odorous. Gas companies need therefore add only minuscule amounts of the smelly chemical to their pipes. Unfortunately for turkey vultures, they also can smell these low concentrations and gather in confusion at leaks.

Turkey vultures are the purifiers of the forest, administering ecological last rites, speeding the material transformation of large animals from carcass to freed nutrients. Their scientific name recognizes this: *Cathartes*, the purger.

The apparently humble role of corpse eater seems to us mightily unpleasant. But the forest is rife with competition for what we disdain. Foxes and raccoons sometimes swipe carrion before vultures can get a mouthful. Black vultures gang up on turkey vultures, their larger cousins, and drive them away from meals. Burying beetles drag away and entomb smaller carcasses.

Mammals, birds, and beetles are rivals, but their competitive importance pales when compared to the microscopic death eaters, bacteria and fungi. From the moment of death these microbes start their work, digesting animals from inside and out. At first, this decomposition helps the vultures by releasing plumes of odor that guide birds down from the sky. Once at the carcass, however, vultures compete with the microbes for the dead animals' nutriment. In hot weather, that competition is won by the microbes within a few days; vultures have to be quick if they are to feed well.

Microbes have more direct methods of competition than mere speed of action. It is no accident that most animals are sickened by a meal of rotting flesh—this sickness is partly caused by poisons that the microbes have secreted to defend their food. "Food poisoning" is impalement on the fence that microbes have erected around their turf. Our tastes have been bent to the microbes' evolutionary wills; we shun rotten food to avoid defensive secretions. Turkey vultures have not been so easily dissuaded. Their guts burn away microbes in battery acid and potent digestive juices. Beyond the gut, vultures have a second line of defense. Unusually large numbers of white blood cells rove their blood, seeking foreign bacteria and other invaders to engulf and destroy. This swarm of defensive cells is kept provisioned by a particularly large spleen.

Turkey vultures' strong constitutions allow them to feed where others would gag or sicken. Paradoxically then, the microbes' toxic barrage benefits vultures to a certain extent by deterring other animals. The line between competition and cooperation is, once again, not so easy to draw.

The vultures' digestive prowess affects the broader forest community. Because vulture digestive tracts are powerful destroyers of bacteria, vultures take their role as purifiers beyond just tidying corpses. Anthrax bacteria and cholera viruses are killed by passage through a vulture. Mammal and insect guts have no such effect. Vultures are therefore unmatched in their ability to cleanse the land of disease. *Cathartes* is truly well named.

Fortunately for those of us who are not fans of anthrax or cholera, turkey vulture populations are stable across most of their North American range. In the Northeast, vulture numbers have even grown, perhaps as a result of increasing densities of deer, all of which must eventually die and be disposed of. There are two exceptions to this good news. Parts of the country that have become dominated by soybeans or other row crops have seen vulture populations decline. Agricultural monocultures support little animal life and have little need for undertakers. Another, more subtle threat lies in deer and rabbit hunters' abandoned or lost kills. Lead ammunition shatters into a fine spray of heavy metal, contaminating shot meat. This is bad for hunters and their families but worse for vultures, which often eat more hunted game than even the most avid shooter. Many turkey vultures are therefore slightly sickened with lead, but the overall population is not in danger from this heavy metal, probably because most vultures have diverse diets that include plenty of nonhunted carrion. In contrast, California condors eat proportionally more lead-peppered bodies than do their cousins the turkey vultures. The few wild-living condors are kept alive by being periodically captured and purged of lead by veterinarians. North American hunting culture necessitates a strange inversion, a purification of the purifiers.

It could be worse. In India, the interaction between technology and vultures has created a much bigger crisis. Widespread use of an anti-inflammatory drug in livestock has inadvertently devastated vulture populations. The drug persists in carcasses and is deadly to the once abundant vultures. Indian vultures are now on the brink of extinction, and as a consequence, putrid dead livestock litters the land. Fly and feral dog populations have exploded, with terrible consequences for public health. Anthrax is common in parts of India. India has the highest incidence of human rabies infections in the world, and most of these are caused by dog bites. The loss of vultures and the subsequent boom in feral dog populations is estimated to cause between three and four thousand extra cases of human rabies per year.

The Parsi community in India has felt the absence of vultures in a different way. Their funerary customs call for the dead to be placed in a Tower of Silence. Corpses are arranged in circles in these squat, open-topped towers where in a few hours vultures turn bodies to bones. Now, with no vultures to consume the dead and with religious proscriptions against burial or fire, the Parsi community is thrown into an extinction-induced philosophical crisis.

India has learned a hard and undeserved lesson about the valuable work of these bald-headed purifiers. The anti-inflammatory drug that caused this woe is now banned in India, but its use continues in some areas, and the vultures have yet to rebound. Regrettably, the same drug is now making inroads in African countries where vultures appear to be just as important and just as vulnerable.

Here in Tennessee, turkey vultures wheeling over the hills are a common sight. So common that it is easy to forget what a gift we have.

September 26th—Migrants

Migrant birds continue to stream over the mandala. Most are traveling south from the boreal forest, a 2.5-million-square-mile expanse of coniferous woodland that stretches from Alaska, through Canada, to Maine. This forest rivals the Amazonian rain forest in size, and it is the breeding ground for billions of songbirds. As the migrants move across the mandala, they carry the resident birds with them in agitated flocks. I watch from a rock ten meters upslope, looking down on surging groups of warblers, chickadees, and downy woodpeckers. The forest is full of their *chip, chek*, and *cheep* sounds: an army of tinkers.

The birds have shed the wariness of the breeding season and come close. Some approach almost within reach of my arm and grant me a clear gaze at their vitality. Their plumage is exquisite. Wing and tail feathers are crisp, crowns are smooth, and body feathers shine as they slide over one another. The birds' late summer molt is complete and every feather is perfect.

For the hooded warblers in the mandala's flock, freshly grown feathers must last a full year. The wear of vegetation, grit, and wind will grind the feathers down, and by midsummer feathers will be ragged-edged and slim. Hooded warblers turn this aging process to their advantage, however. The birds abrade themselves into their breeding costume. Their crowns and throats are muted yellow now, but as the outer edges of these feathers wear away, the black of the breeding plumage is revealed below. This is a thrifty strategy; most other bird species acquire their breeding colors by growing new feathers, each one of which is made from costly protein.

The chickadees, woodpeckers, and hooded warblers grew their fresh fall feathers here, in and around the mandala, following their summer breeding. But most birds in these flocks molted much farther north, in the spruce thickets of Canada. The names of these species, magnolia warbler and Tennessee warbler, belie their ecology. Both were first described and named from migrant "specimens" in the southern states, and this historical peculiarity is fossilized in their names. The magnolia warbler was shot as it fed in a Mississippi magnolia tree; the Tennessee warbler met its fate along the banks of the Cumberland River in Tennessee. Other boreal breeders bear the same historical baggage. Cape May warblers, Nashville warblers, and Connecticut warblers are all birds of the great northern forests. Thus the conventions of zoological nomenclature hide a great truth about the bird life of this continent. The boreal forest is the nursery of North America's avian aristocracy, the warblers, the majority of which nest exclusively or mostly in the north. The mandala is washed twice annually by a tide whose volume and power is born in the land of the wolverine and the lynx.

A distinctly southern sound punctuates the tintinnabulation of boreal birds. A yellow-billed cuckoo clucks from the canopy then bursts into a cascade of hollow *kuks*, drumming out its song. I see the bird high above the mandala, jumping from branch to branch like a monkey. It barely opens its wings as it leaps and cranes its scythelike beak into leaf clusters. It grabs a katydid and gulps the fat insect down before lurching back into the high hidden canopy.

Cuckoos are abundant in the forest around the mandala, but their shyness and their penchant for tall trees mean that they are seldom seen. This bird, like other cuckoos before it, startles me with its strangeness. The cuckoo moves like a primate, sounds like a drummed hollow log, and eats insects that other birds cannot or will not. Its huge beak allows it to swallow big katydids and even small snakes. Caterpillars' defensive hairs deter other birds but not cuckoos. Smooth or hairy, everything goes down the gullet, sometimes with a brisk beaking to snap off hairs, but more often caterpillars are swallowed whole, hairs and all. Cuckoo stomachs are apparently densely matted with caterpillar spines whose barbs lodge in the intestinal wall.

Cuckoos make it their business to break other rules of bird behavior. They don't set up predictable territories but wander nomadically on their breeding grounds looking for clusters of food, then quickly set up camp and breed. The chicks grow rapidly and grow feathers that literally pop open, fully formed. The adults' molt is a casual affair. Instead of shedding and growing feathers in an organized sequence and at a regular time like other birds, cuckoos molt feathers haphazardly, one by one, and spread the molt over their summer and winter grounds. Perhaps psychoactive caterpillar toxins have loosed their allegiance to the status quo or, more likely, their molt strategy is like their breeding style, designed to take advantage of local bursts of richness, then to coast through the lean times. Even their migratory behavior is loose. Ornithologists in South America have captured very young birds, strongly suggesting that some of the "migratory" cuckoos linger and breed on their wintering grounds.

Of all the birds in the mandala today, the cuckoo travels the farthest. The Amazonian forests east of the Andes are its winter home. Most warblers travel slightly less far, to southern Mexico, Central America, and the Caribbean. So the mandala connects, at this moment, nearly the entire New World. Memories of tapirs and toucans brush past thoughts of the tundra's edge; minerals from Ecuador and Haiti fly with sugars from Manitoba and Quebec.

Tonight the warblers will link the mandala outward, beyond the earth's bounds, bringing awareness of the stars into the forest's matter. Having rested and fed all day, the migrants will wing southward in the cool and safety of dark. These flying birds will scan the skies, find Polaris, the North Star, and use its position to head south. The birds gained this astronomical knowledge as youngsters, sitting in the nest, peering into the night and searching for the star that did not swing across the sky. They carry this memory in the wet tangle of their brains, then gaze up in the autumn and steer by the constellations.

Remarkable as it is, knowledge of the stars is a fallible method of orientation. Stars are obscured on cloudy nights, and some first-year birds may grow up in dense forests or overcast regions. Migrant birds therefore have several extra navigational skills. They watch the sunrise and sunset, they learn to follow north-south mountain ranges, and they can detect the invisible lines of the earth's magnetic field.

Migrant birds throw open their senses to the cosmos, integrating sun, stars, and earth as their great tide surges south.

October 5th—Alarm Waves

I sit very still. Time seeps by. A chipmunk walks across the opposite edge of the mandala, barely a meter away. The animal pauses, rummages in the litter with its paws and nose, then disappears into a jumble of rocks. This is a rare encounter. Unlike their suburban or campsite cousins, chipmunks on this mountainside are jumpy creatures. They approach me only after I have sat immobile for a long time. Encouraged by the fruit of my stillness, I settle down and melt into the rock.

Easy breeze. Distant birdsong. The forest's waters are calm. An hour passes.

Then a sharp, hoarse exhalation of air, just a foot or two behind me. I keep still. The deer blasts out another alarm, then a double blast. A flash of white hits my eye and the animal bounds away, snorting as it goes. The deer's alarm belly-flops into the smooth, quiet air, smacking sharp energy through the mandala.

The snorts immediately set three squirrels chattering and whining. Eight chipmunks join in, shooting off rapid *chips*. The wave moves out from the mandala. A wood thrush downslope starts calling, *whippa-whippo-whop*, its head feathers hackled up as it hurls out the call. Distant chipmunks pick up the staccato chorus, carrying it to the edge of earshot.

The deer's alarm at coming so suddenly on an immobile human has reached out hundreds of meters. The agitation, particularly that of the chipmunks, takes more than an hour to recede.

The mandala's birds and mammals live embedded in an acoustic network, each individual connected to others through sound. The forest's news ripples through this network, carrying the latest information about the location and activities of troublemakers. It takes some effort for us urbanized humans to become aware of these traveling signals. We are accustomed to ignoring "background noise," instead taking our cues from the interior noise of our minds. Most of my time sitting or walking in the woods is spent riding waves inside my head, thinking of past or future. I suspect that this is a common experience. Only a repeated act of the will can bring us back to the present, back to our senses.

When we arrive in the acoustic now, we discover that the forest's newsroom is focused on— surprise!—us. We're large, noisy, and fast. And many animals have seen us in our more predatory modes. Those that haven't had personal experience of our guns, traps, and saws quickly learn from their more experienced peers: it is in an animal's interest to pay attention to what alarms others. We are like the hawks, owls, and foxes that seldom get to observe the forest network without triggering noisy news bulletins. Sitting low, staying still, and biding one's time is the only way to slip in. Then we experience the alternating calm and clatter of the news wires. Hikers, for example, are preceded by bow waves that arrive minutes before their chatter and laughter. More minor disturbances, such as a branch falling or the overflight of a crow, send quieter and more short-lived pulses through the network. The deer's alarm at stumbling upon me was, on the other hand, a surge, a bold headline.

Tuning in to the network has clear advantages for the forest's animals. Awareness of potential danger gives listeners a head start in deciding how to react. But the advantages of actively contributing to the waves of information are not so obvious. Why call when you see a predator? Why not eavesdrop on others but keep mum? Calling attention to yourself by making a loud sound when a predator approaches seems nonsensical.

For animals with kin nearby, the costs of alarm calling may be outweighed by the need to protect family. Although it is late in the season, some chipmunks and squirrels around the mandala have youngsters with them, so their squeaks and trills give advanced warning to their offspring. But many animals use alarm calls when family are not present, so other benefits must also be in play. Some alarm signals are designed to actively communicate with the predator, drawing attention to the animal in the moment of danger. In doing so they may reap the paradoxical benefit of telling the predator who and where they are. From the predator's perspective, prey that have seen your approach and are poised to flee are likely to be hard to catch. The predator's time would be better spent looking for unwary prey. Alarm calling can therefore provide a direct benefit to the caller by advertising the unprofitability of an attack and thus buying safety. "I've seen you—you can't catch me. Move on."

White-tailed deer take this advertisement one step further. As they run from predators they pump their tails up and down, flashing a white rump and undertail at the pursuer. Their run is interspersed with long upward leaps, losing time that could be spent hoofing forward. Those flashing, prancing displays must have a function beyond telling the predator that it has been seen: running away is already a clear signal that the deer has detected the predator. It is possible that the deer is communicating its vigor and hence ability to escape. Only healthy deer can afford to punctuate their run with wasteful flourishes; weak or sick deer cannot risk their lives with time-wasting bounding displays. This idea has not been thoroughly tested in white-tailed deer, but similar puzzlingly exaggerated displays in gazelles do seem to be honest signals of condition.

The animals' acoustic network has an invisible analog among the forest's plants. When insects chew on leaves, they trigger a physiological reaction from the host plant that not only deters further damage to the host but also alerts neighboring plants. Damaged leaves turn on genes that produce a flush of chemicals. Some of the defensive chemicals evaporate and perfume the air around wounded plants. The wet interior of the neighboring plants' leaves is soaked in these molecules and, like aroma in a human nose, the molecules dissolve and move into the surrounding cells. Here the chemicals turn on some of the same genes that produced defensive chemicals in the original plant. Unwounded plants around a damaged colleague therefore become less palatable to insects. The trees are listening.

When I sit or walk in the forest I am not a "subject" observing "objects." I enter the mandala and am caught up in webs of communication, networks of relationships. Whether or not I am aware of it, I change these webs by alarming a deer, startling a chipmunk, or stepping on a living leaf. Dissociated observation is not possible in the mandala.

The webs change me also. Every inhalation carries hundreds of airborne molecules into my body. These molecules are the aroma of the woods, the combined fragrance of thousands of creatures. Some aromas are so pleasing to humans that we have domesticated them, extracting "perfumes." At least one such perfume, jasmonate, is an alarm chemical, communicating danger among plants. Perhaps our olfactory aesthetics reflect a desire to be wedded to nature's struggle?

But perfumes are the exception. Most of the forest's molecules bypass my sense of smell and dissolve directly into my blood, entering my body and mind below the level of consciousness. The effects of our chemical interpenetration with plant aromas are largely unstudied. Western science hasn't stooped to take seriously the possibility that the forest, or the lack of it, might be part of our being. Yet forest lovers know very well that trees affect our minds. The Japanese have named this knowledge and turned it into a practice, *shinrin-yoku*, or bathing in forest air. It seems that participation in the mandala's community of information may bring us a measure of well-being at the wet chemical core of ourselves.

October 14th—Samara

The forest's colors are gradually turning. The spicebush in the mandala is mostly green, but a few leaves are flecked with yellow. Color in the ash next to the spicebush is faded, and the outer leaves are drying and bleached. Above me, the maple and hickory still show their summer colors, but the leaves of a large hickory upslope have all turned tan and gold. A few scattered leaves have fallen, refreshing the top surface of the leaf litter and giving a quiet crunch to moving animals.

A winged maple seed flashes past my face. It whirs in a blur of light, like a flying knife at the circus. The seed helicopters down, then strikes the leaf of a toothwort herb, falls between two dead leaves on the forest floor, misses a sandstone pebble, and lodges vane up, seed down in a crevice in the humus. A fine place to germinate—this was a lucky fall.

April's maple flowers have finally ripened and, after months of slow growth, helicopters are scattered all across the forest floor. A few nestle in dark openings in the litter, but most are exposed on the dry surfaces of leaves or rocks. For all the whirling drama of their flight from the canopy, the ultimate fate of the maple seeds is determined by the particularities of where they land. Rough surfaces excel at catching windblown seeds, so mossy rocks snare more seeds than do bare rocks. The leeward sides of trees accumulate more seeds than the windward. Animal predators destroy seeds by eating them, or inadvertently spread and sow them by storing them for a meal that never comes because of forgetfulness or death.

There is little that wind-dispersed seeds can do to select a premier germination spot. They don't get carried to a fertile ant nest like *Hepatica* seeds, or deposited like cherry seeds in a pile of manure, or wiped onto a convenient branch by the beak of a mistletoe-carrying bird. But the maple seed's helplessness about the choice of its final destination does not mean that the seed has no power. The seed's skill comes before the final landing.

This morning, no seeds were falling on the mandala. Now, in the late afternoon, they rain so densely that their crackling impact on the ground sounds like a forest fire. This is no coincidence. The thin strip of tissue that holds the seed to its parent is weakest on dry afternoons. These afternoons are also when the wind is strongest—trees time the release of seeds to catch the best wind. Of course, the tree has no central air traffic controller to tell seeds when to depart. Instead, the materials used to fasten the seed to the mother tree, as well as the shape and strength of the attachment, determine when and how seeds will be released. Millions of years of natural selection have tuned the design of these release mechanisms.

There is more to the trees' strategy than simply dumping seeds into dry air. Flying seeds have two roads ahead of them. The "low road" carries them down from the canopy to the forest floor around the parent. These seeds travel, at most, a hundred meters or so from home. The "high road" takes seeds above the canopy, into the open sky where they may travel miles.

Few seeds take the gravity-defying high road, but it is of great importance to the fate of tree species. Rare long-distance dispersers have a strong effect on the genetic structure of species, on the ability of species to persist in fragmented landscapes, and on how fast species move in response to retreating ice ages or advancing global warming. Like human history, the narratives of ecology and evolution hinge on the actions of a few individuals that travel across continents and settle far from home.

Maples try to buy a ticket on the *Mayflower* by contriving to launch seeds into gusty updrafts. They do so by preferentially releasing seeds in the upward-blowing air of eddies and gusts but holding tight in downdrafts. Many wind-dispersed trees concentrate their seeds at the top of the canopy, increasing the chances of launching a seed that can catch an updraft. Maples in the mandala have an extra advantage. The prevailing winds blow unobstructed across the valley below and are then deflected upward by the steep slope on which the mandala sits. The wind therefore gives the mandala's seeds an extra skyward puff in their fight against gravity.

Each tree casts a "seed shadow" that is darkest, most densely packed with seeds, in the tree's immediate surroundings but theoretically extends across the entire continent. A glance upward confirms that the maple seeds that flutter onto the mandala are almost all the stay-at-home variety, having fallen from trees within easy gliding distance. Mixed among them are a very small number of competitors from other parts of this forest, or perhaps a rare seed that traveled here on a thermal updraft, like a vulture, from tens or hundreds of miles away.

The long reach of seed shadows makes the study of seed dispersal difficult. It is easy to gather information about the vast majority of seeds that stay close to their parents. But the offspring that get flung into the open sky are nearly impossible to track—yet they are the key players in the grand story of each species' history.

Lacking a drone spy plane with which to track soaring seeds, I turn my attention to the maple seeds on the mandala's surface. The diversity of forms is remarkable, particularly in the shapes of the wings. Some have three times the surface area of others. A few are ruler-straight; others curve downward like a boomerang, and others arch upward. On most seeds, a notch indents the wing where it meets the seed, but some are notchless. The angle and depth of the notch varies, as does the fatness of the wing. The mandala is a botanical air show, with every airplane wing shape on display, and a few shapes that no human engineer would dare use.

The diversity of shapes causes the maple seeds to fall with very different styles. The most obvious seeds are those that don't fly but plummet. One in five seeds lands joined to its sibling. These pairs don't spin at all but plunge and smack into the ground under the tree. Singletons with small or hunched wings also drop without spinning. These are exceptions, however. Most seeds fall freely for a second or two, and then start to spin. The wing rotates so that its rib, the fatter edge, slices through the air with the wing's thin vein following. This spinning airfoil generates lift, slowing the fall. A drifting seed can obviously glide farther from the parent than can a seed that falls like a stone. But the extra time in the air also increases the likelihood that a turbulent updraft will cause it to float upward. Whether through shallow descent or lucky ascent, the wind smears the trees' seed shadows outward, reducing competition among siblings and dropping hopeful progeny across a wide area.

Botanists call seeds that produce their own lift samaras. Technically, a samara is not a seed but a special fruit, formed by the mother's tissues, holding the seed inside. Ashes and tuliptrees also produce samaras, although neither generates the same amount of lift as the maple's spinning blade. The maple's asymmetry gives it an advantage. Its samara is designed to flow through the air like a bird or an airplane wing, with a slicing leading edge. Ash and tuliptree samaras are symmetrical and cannot achieve the maple's elegant spin. Instead, they twist rapidly about their long axes as they fall, preventing the wing from biting into the air. These species rely less on their own airfoil and more on the strength of the wind to carry their seeds. Accordingly, ashes and tuliptrees cling tightly to their samaras, letting them go only when the wind rages hard.

Maple samaras live in a little-known border country between the aerodynamics of fast, large objects, like cars and airplanes, and the aerodynamics of slow, minuscule objects, like motes of dust. Airplanes experience their surroundings as relatively free of friction, but dust flecks are so small that friction is about all that they experience. In other words, as an object gets smaller, its world comes to resemble a jar of cold molasses: hard to swim through but easy to float in. The size and speed of samaras puts them in low-grade maple syrup, perhaps appropriately. Engineers have shown that this syrupy air forms a swirl over the leading edge of a rotating blade. This miniature vortex sucks on the upper surface of the spinning samara, slowing its descent.

The aerodynamic consequences of the variety of maple samara shapes are hard to assess. But students of maple samaras have tossed seeds from balconies and drawn two general conclusions. First, wide wingtips produce turbulence and likely slow the wing's spin, reducing lift. Curved wings are likewise less able to generate lift than straight wings. So, fat-tipped, curvy samaras are poor fliers in the ordered air of a laboratory building. But most of the seeds in the mandala have fat tips and curves. Are the samaras defective versions of an elusive perfect form? Or do maples know something that we do not about the advantages of fat-tipped, curved, or otherwise "deficient" wings?

Wind in a forest is a confusion of swirls and puffs. The forms of the samaras seem to me to be botanical incarnations of the wind's complexity: a wing for every eddy, a curve for every gust. This diversity of biological form is not restricted to samaras but is a general theme in the forest. Close examination of almost any structure here—leaves, animal limbs, twigs, or insect wings—reveals variability everywhere. Some of this irregularity comes from the different environmental contexts that individuals find themselves in, but much of it has deeper roots in genetics, made possible by the reshuffling of DNA in sexual reproduction.

The fact that individuals differ subtly from one another seems a minor detail of natural history, but such variability is the basis of all evolutionary change. Without diversity, there can be no natural selection and no adaptation, as Darwin knew when he devoted the first two chapters of *On the Origin of Species* to variation. The samaras' diversity therefore points indirectly to the invisible workings of evolution. From these varied forms will be chosen the next generation of maples, specially adapted to the winds blowing through this mandala.

October 29th—Faces

Squalling bands of cold rain scoured the forest last week, bringing to ground the first significant accumulation of leaves. Now a strong sun has baked the leaf litter, and every moving animal stirs up loud rustling. Crickets and katydids are quickened by the warmth, and they sing with vigor: regular high pulses from crickets hiding beneath downed leaves and raspy trills from angle-winged katydids clinging under branches. Unlike the dawn chorus of birds in the springtime, the fall-breeding crickets are loudest in the midafternoon, when their bodies have sponged the day's heat.

The precise songs of insects are punctuated by untidy crackling sounds. A gray squirrel shambles toward the mandala, intermittently plunging its nose into the litter. The squirrel seems fevered, its body trembling with disorganized energy. The animal continues alternate surges and rummaging until it reaches a tree, where it scrabbles up and disappears from view. A few minutes later it descends, headfirst, with a hickory nut in its mouth. The squirrel catches me with its dark eyes, then freezes. Its head is held tipped up, and the tail straightens parallel to the tree trunk. The squirrel watches. Then trembling waves agitate the tail. The fur on the tail flattens, turning a brush into an undulating fan.

I hear quiet drumming as the tail pulses. Somehow, the flattened tail is substantial enough to beat out a warning tattoo on the trunk. I have seen the tail-fluttering display many times but have never been close or quiet enough to hear the subtle tapping. This novelty is not just a consequence of my weak powers of observation—I am likely not the intended recipient of the signal. Gentle drumming sounds carry poorly through air, but the vibrations move with great efficacy through wood. Other squirrels in this tree, especially those in tree holes, will hear the warning through both their ears and their feet.

The squirrel completes its descent in spurts, alternating stationary drumming with darting rushes down the trunk. It reaches the ground, runs to the far side of the trunk and, after poking its head out from behind the tree to glance at me one last time, bounds away, the hickory prize locked in its jaws.

The drumming squirrel was not alone. Within five meters of me at least four other squirrels ply the mat of leaves, with more above in the branches. The hickory adjacent to the mandala is one of the few trees in this patch of forest with nuts still falling, making it a popular destination for squirrels, whose winter survival depends on body fat and nut stores. Competition among the foragers stirs up a frenzy of crinkling leaves and chittering mouths.

I sit and listen as the afternoon turns to evening. The squirrels' urgent sounds rise and fall against the backdrop of the constant, mellow trilling of crickets. As light starts to dim, a new sound pushes into my awareness. The sound comes from behind me, upslope. I am loath to swing around and startle whatever animal is making the unfamiliar noise, so I sit motionless and focus on the sound. Unlike the nose-pushing or bounding of squirrels, this sound is steady, a continuous rustle, getting louder, like a large ball rolling through the litter. The strange sound builds. It is headed directly at me, and I feel a small surge of anxiety. Slowly, I twist just my neck, hoping to steal a glance.

Twelve paws paddle the litter as three raccoons trundle toward me. Their movement is focused, calm, and purposeful. They seem to glide down the hill, like mammalian caterpillars, fuzzed in silvery gray. They are slightly smaller than the adults that I see in these parts; perhaps they are young of the year, born this spring.

I sit directly on the line of the raccoons' path, and they come within a foot before abruptly stopping. My neck is turned the wrong way, so the animals have moved out of sight. I pour my attention into my ears. The raccoons puff and sniff as they stand, making a nasal investigation. After half a minute, one snorts gently, giving a soft, fleshy oink. At this, all three continue on their path, skirting me by a foot or two. They show no sign of alarm as they come into sight, then pour away down the hill.

My first reaction to the raccoons was that of surprise, a jolt of excitement as the strange sound resolved into the advancing trio. Then the raccoons' appealing faces came close: dark velvet masks set in crisp white borders, obsidian eyes, rounded ears perked jauntily, and slender noses. All this set in ruffs of silver fur. One thing was evident: these animals were adorable.

My zoological self was immediately embarrassed by these thoughts. Naturalists are meant to have outgrown such judgments. "Cute" is for children and amateurs, especially when applied to a common animal like a raccoon. I try to see animals for what they are, independent beings, not as projections of desires leaping unbidden from my psyche. But, like it or not, the feelings were there. I wanted to pick up a raccoon and tickle it under the chin. Surely this was the ultimate humiliation of the zoologist's scientific hauteur.

Darwin might have sympathized with my plight; he knew the emotional power of faces. In *The Expression of the Emotions in Man and Animals*, published a decade after *On the Origin of Species*, Darwin explained how human and animal faces reflect emotional states. Our nervous systems scribe our inward feelings onto our faces, even when our intellect would rather conceal what is within. Sensitivity to the nuances of facial expression is a core part of our being, Darwin claimed.

Darwin focused on the nervous and muscular mechanisms that translated emotions into facial expressions, implicitly assuming that observers of faces would interpret them correctly. In the early and mid-twentieth century, Konrad Lorenz, one of the first proponents of the evolutionary study of animal behavior, made explicit what Darwin had assumed. Lorenz analyzed faces as forms of communication, analyzing the evolutionary benefits that animals might gain from being sensitive to facial expressions. Lorenz also extended Darwin's analysis by considering why humans are attracted to some animal faces and not to others.

He concluded that our affinity for the faces of human babies could mislead us when we viewed animals. We see baby-faced animals as "lovable," even if the animals' true characters are decidedly not cuddly. Lorenz believed that large eyes, rounded features, proportionally large heads, and short limbs all release in us an instinct to embrace and to pet. Misplaced feelings also apply to other facial types. Camels hold their noses above the level of their eyes, causing us to view them as haughty and disdainful. Eagles have resolute brow ridges and mouths set in narrow, determined lines; we see in their faces leadership, imperialism, and war.

Lorenz's view was that our perceptions of animals are strongly discolored by the rules that we use to judge human faces. I suspect that he was right, but only partly so. Humans have interacted with animals for millions of years. Surely we might have picked up the ability to see that a raccoon isn't a baby? This ability would have served us well. Any ancestor who could correctly interpret the danger or utility of other animal species presumably had an advantage over those with no zoological acumen. I suspect that our unconscious reactions to animals are shaped by these judgments as much as they are by misapplication of rules evolved for human faces. We have an affinity for animals that pose little physical danger to us: those with small bodies, delicate jaws, and averted, submissive eyes. We fear those whose eyes stare us down, whose faces bulge with jaw muscles, and whose limbs could outrun and overpower us. Domestication is the latest chapter of our long evolutionary relationship with other animals. Those humans who could work effectively with animal partners gained hunting dogs, goats for meat and milk, and oxen for labor. Agrarianism requires a sophisticated ability to read other animals.

When the raccoons ambled into view, my ancestors called to me through the evolved intricacies of my brain: "Short legs and delicate jaws, squat bodies; these fellows pose little risk. The body looks well muscled, a decent meal; they show no fear, perhaps it would be fun to keep one; charming faces, like little babies." All this wells up wordlessly from the past and suffuses me with an attraction to the animals. Later, the words try to explain the longing, but the process of attraction happens first entirely below the level of reason, layered under words and language.

Perhaps I should not have felt so embarrassed at my immediate and strong attraction. What I interpreted as the humiliation of my pretensions as a zoological sophisticate was in fact an education in my own animal nature. *Homo sapiens* is a face-reading species. We ride waves of emotional judgment all our lives, drawing rapid, unconscious conclusions every time we see a face. The raccoons' faces gave me a psychologically incongruous shock, waking my conscious mind into discomfiture. But my reaction to the raccoons was just an extension of the responses that I experience dozens or hundreds of times each day.

As the raccoons walk away, crunching over the dry litter, I sense that my observation of the forest has held up a mirror to my own nature, a mirror that is less clouded here than in the synthetic modern world. My ancestors lived in community with animals of forests and grasslands for hundreds of thousands of years. As in all other species, my brain and my psychological affinities have been built by these millennia of ecological interactions. Human culture now modifies, blends, and transforms my mental predispositions, but it does not replace them. By my returning to the forest, albeit as an observer rather than a full participant in the community, my psychological inheritance starts to reveal itself.

November 5th—Light

The sound of my footsteps has changed radically this week. Two days ago, the forest floor was deep with sun-dried fallen leaves. Silent movement was impossible; walking was like traversing a field of balled crinkle-wrap. Today, the crash and crunch of autumn's shed leaves are gone. Rain has relaxed the leaves' tense curls, and animals move across the wetted, muted ground with silent steps.

The rain followed a weeklong dry spell, so the moisture-loving animals of the leaf litter are moving to the surface after many days of hiding. The most striking of these small animals is a slug that glides over a patch of emerald moss. Although I have seen these slugs in other parts of the forest, this is my first glimpse of one in the mandala and my first view of one traveling exposed in the midafternoon. Unlike the European slugs that plague gardens in our region, this species is a native and lives only in its aboriginal forest habitat.

The familiar European slug has a saddle mounted on its back, just behind the head. This smooth patch of skin is the mantle that covers the lungs and reproductive organs. The native slug in the mandala belongs to the Philomycid family, all members of which have distinctive mantles that stretch the entire length of the back, like the icing on a pastry éclair. Philomycids therefore look more decently attired than their European cousins that have an unpleasant naked aspect. The expanded mantle also provides a canvas for beautiful markings. The slug in the mandala has a matte silvery ground color with dark chocolate decorations—a thin line scribed along the center of the back and fingers reaching from the mantle's edge to the midline.

On the rain-freshened green moss, the slug's markings are striking and richly contrasted. As the slug slides onto the lichen-covered rock face, the effect changes. Color and form melt into the variegated surface; beauty remains, but it is the camouflaged beauty of belonging.

My focus on the slug is interrupted by the sound of heavy rain against the tree canopy. Distractedly, I pull on my rain jacket, keeping my eye on the slug. But I was fooled: there is no rain, just pelting falls of wind-thrown leaves. The squall of leaves settles, adding another stratum to the thickening bed over the mandala. Most of this bed was deposited in the last couple of days, the leaves' tenacious hold on their twigs broken by the extra weight of moisture from the rain. Two days ago, the forest canopy was thickly metaled with the bronze and gold of hickory and maple leaves. Today, a few sparse scraps hold out, but the canopy's armor is gone.

Finally, rain comes, starting with big, cold, splatting drops and settling into an even shower. More leaves descend. A tree frog rasps loudly from an oak trunk, greeting the rain with four bursts of song. Crickets fall silent. The slug continues its exploration, at home in the slick air.

I huddle in my rain jacket, feeling an unexpected sense of aesthetic relief at the changing forest. This is hardly reasonable—autumn rains presage the cold and the pinching in of life for winter. But some quality that was missing from summer has returned. As I gaze through the rain, I realize that I am buoyed by an expanded quality of light under the opened forest canopy. My view of the forest seems deeper, fuller. I am released from a narrowness of luminosity that I hadn't known existed.

The herbs in the mandala appear to sense the change also. Sweet cicely plants that grew in late spring, then faded through summer, have put out fresh sprays of growth. Each plant has several new sets of lacy leaves. Presumably these low-growing herbs are grabbing a few days of extra photosynthesis under the thinned canopy. Despite the short days, enough light now reaches the mandala's surface to make worthwhile the investment in new growth.

Without an umbrella of leaves, the forest floor is brighter. But my response, and I suspect part of the herb's response, is due as much to the shape or quality of the light as it is to increased intensity. The loss of leaves has widened the light spectrum and freed the forest's painterly hand.

Summer light is constrained, clipped back into a narrow range. In the deep forest shade, yellow-green light reigns; blues, reds, purples are all muted, as are the hues that form in combination with these missing colors. Sunlight that lances through the canopy is dominated by intense orange-yellow, but these beams are so narrow that the sky's blue or cloudy white are missing. Near larger gaps in the canopy, the ferny colors of the shade are augmented by indirect colors from the sky, but the sun's copper seldom reaches through. Life under the summer canopy is played under a miserly range of stage lights.

Now reds, purples, blues, oranges can mix in thousands of tones and hues: ash skies, sand and saffron leaves, blue-green lichens, silver and sepia slugs, and tree limbs in dun, russet, and slate. The forest's National Gallery has unlocked its collections. After living a season swimming in the yellow and green light of Van Gogh's sunflowers and Monet's lily pond—masterly works, but only a small portion of the full collection—we're allowed to roam the galleries reveling in the full depth and range of visual experience.

My strong unconscious relief at the changed forest light suggests something about our visual sense. We crave rich variegations of light. Too much time in one ambience, and we long for something new. Perhaps this explains the sensory ennui of those who live under unchanging skies. The monotony of blank sunny skies or of an endless cloud ceiling deprives us of the visual diversity we desire.

. . .

The mandala's light environment affects far more than my human sense of aesthetics. Plants' growth is mediated by light, as are the feeding and breeding of most animals. Sensitivity to variations in illumination is therefore a central part of the lives of forest creatures. Herbs living on the forest floor in autumn grow and embrace the wavelengths of light that were formerly intercepted by tree leaves. Limbs on trees use the strength and color of light to direct their growth into sunny openings and away from other limbs. Inside plant cells, light-harvesting molecules respond to changing light minute by minute, assembling then disassembling as needed.

Animals also modulate their behavior as light changes. Some spiders adjust the color of their web silk to the particularities of brightness and color in different parts of the forest. Tree frogs melt into their background by moving pigments up and down within their skin, tuning the brightness and color of the skin to the surface on which the frogs find themselves. Displaying birds position themselves in the light environments that best show their feather colors.

Birds with red plumage have a particularly rich set of opportunities in and under the forest canopy. Birds such as cardinals and scarlet tanagers seem gaudy and conspicuous when we see them isolated on the page of a field guide. Yet the red part of the light spectrum is weak in the green gloom of forests. A "bright" red bird appears dusky and dull in the forest shade. But when the bird steps into a patch of direct sunlight, colors blaze and feathers dazzle. By hopping into and out of sunflecks, red forest birds can transform themselves from sulkers to show-offs, then back again, all in an instant. In my experience, woodpeckers are particularly adept at this trick. All seven woodpecker species here have red crests or crowns, and all are experts at manipulating the light. When woodpeckers are feeding quietly they are devilish to locate, but when they are advertising their ownership of a patch of forest, or displaying to a mate, they are like burning torches at dusk, unmissable.

Conspicuous displays are impressive, but they are not the most masterly of the animals' adaptations to light. Obscurity is much harder to achieve. Not only must a camouflaged animal match the hue and tone of its environment, but the rhythm and scale of the textures on its surface must echo those of the background. Any deviation from the visual properties of the surroundings produces visual dissonance and a potential failure of disguise. There are thousands of ways to stand out in the forest but only a few ways to blend in.

The evolution of camouflage is a finicky process in which the particularities of place matter a great deal. Therefore, animal species whose lives are played against just one visual backdrop, moths that rest only on hickory bark for example, are more likely to evolve camouflage than species that move among backdrops, such as moths that flit from hickory bark to maple twigs to spicebush leaves. Mobile animal species rely on other forms of defense—fast escape, noxious chemicals, and protective spines.

For camouflaged species, concealment within a particular microhabitat is a great short-term adaptation. In the longer term, such specialization can be a trap; the fate of the camouflaged species is tied to the background on which it rests. Moths that are beautifully disguised on hickory bark thrive as long as hickories are plentiful, but if hickories should decline, these otherwise poorly defended moths will be picked off by sharp-eyed birds in new visual environments. Even if hickories continue to be abundant, moths that specialize on hickory bark are ecologically constrained and are less likely to evolve new ways of living in the world. Their cousins that rely on other methods of defense can explore new habitats without incurring the severe penalties that failed camouflage entails. In some ways, therefore, the textbook example of camouflage evolution—English peppered moths evolving dark wings as trees in their environment changed from unpolluted gray to sooty black—is unrepresentative of the evolutionary pressures experienced by moths. Seldom does a lucky mutation allow a camouflaged animal to switch backgrounds so easily. The complexity of the visual environment and the sophistication of predator eyes make the evolution of camouflage more fraught and constrained than the textbooks suggest.

The colors on the slug meandering across the mandala match the colors of the lichens and wet leaves through which the animal moves. This straightforward form of camouflage is augmented by an additional layer of visual trickery. The irregular flames of dark pigment that lick up from the mantle's edge serve to break up the outline of the slug. These disruptive patterns deceive the eye by creating the perception of edge where there is no edge, thus distracting the neural processors in predators' eyes and brains, concealing the real edge with seemingly meaningless patterns. This decoying of pattern-recognition systems is surprisingly effective. Experiments with feeding birds show that disruptive patterns, even when the patterns are formed by conspicuous colors, can match or beat the performance of simple color-matching camouflage.

Disruptive patterns do not depend on an exact match between the animal and the color and texture of its background. Animals with disruptive patterns therefore stay hidden against many different backgrounds, avoiding the constraints imposed on animals with camouflage that perfectly matches just one habitat. The slug remains protected on green moss, even though the slug has no green on its skin. Its fake edges give no hint of the true, edible shape of the animal. Only a prolonged gaze unmasks the deception. Scanning predators cannot afford to sit watching one small patch of moss for an hour or more, as I have done.

Predators are not without countermeasures. Some peculiarities in our human visual ecology may be partly explained by the visual sparring between predator and prey. Military planners in the Second World War noticed that color-blind soldiers were better at seeing through camouflage than were soldiers with normal vision. More recent experiments have confirmed that dichromats (people with two types of color receptors in their eyes, so-called red-green color-blind) are better camouflage breakers than are trichromats (people with three receptor types, the more common condition in humans). Dichromats detect boundaries in texture that are missed by trichromats, who are fixated on and misled by variations in color.

The superior pattern-finding abilities of dichromats may seem a peculiar but unimportant quirk of an unlucky mutation. Two facts suggest otherwise. First, the frequency of dichromatism in humans, two to eight percent of all males (the genetic change is on the male sex chromosome), is much higher than would be expected if the condition were a maladaptation. Such commonness suggests that evolution may, in some circumstances, smile on the condition. Second, our cousins the monkeys, specifically the New World monkeys, also have both dichromats and trichromats living together within the same species. Dichromats in these species make up half or more of the population, again suggesting that dichromatism is not just an accidental defect. Feeding experiments with marmosets found that dichromats have an advantage over trichromats when light is dim, perhaps by seeing patterns and texture that were missed by trichromats. In bright light, the advantage is reversed; trichromats can find ripe red fruit faster than can dichromats. The diversity of these monkeys' ways of seeing may therefore be a reflection of the diversity of light conditions in the forest.

New World monkeys generally live in cooperative groups, so it is to everyone's advantage to have both types of vision in the same group—food can be found in all kinds of lighting conditions. Whether the same explanation applies to humans is not known. We also evolved in a social context of extended groups, so it is possible that dichromatism now exists in humans because of past natural selection. Perhaps groups with some dichromats fared better than exclusively trichromatic groups, thus passing the genetic propensity for dichromatism to future generations. These are interesting speculations, but no one has tested them by examining human visual performance under conditions that approximate those of our ancestors.

. . .

My response to the changing light in the forest was subconscious and manifested itself in my aesthetic sense. It is tempting to brush away such aesthetic responses as merely human inventions, unconnected to the forest. What could be less natural than the overcultivated tastes of a human? But it turns out that our human aesthetic faculties do reflect the forest's ecology. Our sensitivity to the tone, hue, and intensity of light is bound up with our evolutionary heritage. Even the diversity of our visual abilities may echo our ancestors' ecology.

We live in a civilized world where light is usually as unsubtle as a flashing computer screen or billboard. The mandala's changing autumn light stirred in me an awareness of the forest's more subtle illumination. I am coming late to this awareness. Sweet cicely knew about the autumnal light weeks before I did and has unfolded new leaves. Many generations of natural selection taught the slug about light and scribed markings onto its mantle. Spiders, cardinals, woodpeckers, and frogs all know how the forest is lit and tune their behavior, their silks, feathers, and skins to the forest's rich light. As rain brings down the last golden leaves, I also start to see.

November 15th—Sharp-shinned Hawk

We have crossed a seasonal threshold. Ice has returned to the mandala, coating the leaves of low-growing herbs with a fuzz of crystals. Frosts have brushed the canopy intermittently for a week or so, but this is the first autumn freeze to have reached the ground. Unlike deciduous trees that drop their leaves to avoid freeze damage, many herbs persist through the cold, loading their cells with sugars to act as antifreeze. They also suffuse their leaves with purple pigment that protects the cells, shielding them from sun damage when the cells' usual light-absorbing machinery is iced up. Herbs that were formerly entirely green, *Hepatica* and leafcup, are now edged with deep purple, the mark of approaching winter. These leaves will hold on through the entire winter, eking out small amounts of photosynthesis on warm days and dying back only when fresh springtime growth supersedes them.

Although mornings are frosty, there is still plenty of animal life in the mandala. As the day's temperature climbs, small insects swarm into the air, and the leaf litter is populated by ants, millipedes, and spiders. These invertebrates are a rich source of nourishment for birds, some of which have recently arrived here, fugitives from the snowstorms that have choked off their food supply in more northerly forests. One of these birds, a winter wren, visits as I sit at the mandala. It lands next to me, poking its needle beak into the folds of my bag and the hem of my jacket before darting to a viburnum shrub. There, it hangs from a twig, regarding me with one black eye from a tilted head, then takes to the air and flies into a tangle of downed branches a few meters away. Its tiny dusky body disappears into the thicket, moving more like a mouse than a bird. The chattering calls of these wrens have been common for at least a week, but I feel lucky to have been investigated so closely by this bird; they are usually much more wary.

Unlike the migrant warblers that have left the mandala and are now in Central and South America, these wrens make a relatively short journey, lingering in North American forests all winter. In most years, this is a successful strategy, saving a costly transcontinental flight and allowing the birds to get back to their breeding ground swiftly. But the winter wrens' preference for feeding on the ground and on fallen wood makes them vulnerable to hard winters. The combination of cold and deep snow in southerly forests can cause drastic die-offs in some years.

This visit from an inquisitive wren was my second unusual bird encounter of the day. On my walk into the forest, a flash of cobalt blue sprung vertically from the center of the mandala. The sharp-shinned hawk's wings and tail splayed out as she curtailed her swoop and, in a blink, she rose twenty feet, bouncing off air. The wings twisted, the body leveled, and the bird swung in an ascending arc, alighting on a maple branch. She sat briefly, holding her back and long tail vertically, then slipped down the slope with her wings and tail held in a motionless T.

Like a pebble skimmed over ice, the bird's motion seems effortless and smooth. As she slid out of sight into the haze of trees I felt gravity like a restraining strap, leashing me to earth. I am rocklike, a clumsy boulder.

The hawk's mastery relies on a careful proportionality between weight and power. The sharp-shinned hawk likely weighs just two hundred grams, several hundred times lighter than me. Her pectoral muscles are several centimeters thick, meatier than many human chests and forming one-sixth of the bird's weight. A contraction of the hawk's muscles therefore sends her soaring, like a beach ball launched skyward by the kick of a powerful leg.

Humans have tried to follow the hawks, but medieval tower-jumpers and Haight-Ashbury trippers were all given the same hard, unyielding answer to their requests for airy freedom. Only by transcending our bodies' limitations with a slug of power from fossil fuel have we been able to break the strap that holds us to earth. To do so under our own power would require grotesque modifications of our bodies: either chest muscles six feet deep, or an impossible reduction of the bulk of the rest of our body. We are just too puny for our leaden frame. The story of Icarus's flight from Crete might therefore be instructive about the dangers of hubris but is a poor teacher of aerodynamics. Gravity would have taught him humility long before the sun could deliver its melting judgment on his wings of wax and feathers.

The balance between weight and power is the background to the rest of bird biology. Land-bound animals carry their reproductive organs around all year, but birds atrophy their testes and ovaries after they breed, shrinking them to tiny specks of tissue. Teeth are likewise dispensed with in favor of a paper-thin beak and grinding stomach. The droppings on our car windshields are another part of the birds' strategy. By excreting white crystals of uric acid instead of watery urea, birds forgo the need for a heavy bladder.

Bird bodies are only partly solid. Much of the body is filled with air sacs, and many of the bones are hollow. These tubular bones provided humans with an unexpected gift. Archaeologists in China have uncovered nine-thousand-year-old flutes made from the wing bones of red-crowned cranes. The flute maker bored holes into the bone, creating a scale similar to the modern Western "do, re, mi." Neolithic artists were therefore able to turn the magic of flight into another wind-borne delight.

The bubble-wrap lightness of the hawk is given an extra upward boost by the physiology of her thick pectoral muscles. Because bird bodies run hot, at temperatures above forty degrees Celsius, the molecules that make up muscles react with speed and vigor, doubling the strength of their muscular contractions compared to our languid mammalian squeezes. Bird muscles are netted with capillaries that carry blood from a heart that is proportionally double the size of a mammalian heart and is far more efficient than the leaky pump that the birds' ancestors, the reptiles, possessed. Blood is kept oxygenated by the bird's unique one-way lung that uses the air sacs in the rest of the body as bellows to keep air streaming across the lung's wet surface.

All this impressive physiology produces more than mere flight. The hawk dances on air. In just ten seconds, she stopped a rapid dive, rose vertically while turning, swept in a new direction, flapped upward, and curved into a rising arc, ending with a stall that parked her feet directly over a maple branch. The precision and beauty of bird flight is so familiar that our wonder is jaded. We should be frozen in amazement at the cardinal landing on the feeder or the sparrow banking around cars in the parking lot. Instead, we walk by as if an animal pirouetting on air were unremarkable, even mundane. The hawk's dramatic rise over the mandala's center jolts me out of this dullness, pulling away the blinding layers of familiarity.

Because the wing bones of birds are built on the same design as our forearms, we can imagine, at least in part, the lifting and folding of bird wings. But feathers add an alien layer of refinement, one that eludes our intuitive understanding. Hair is the closest human analog, but our simple protein ropes are slack and lifeless compared to the intricacy and control of bird feathers. Each feather is a fan of interlocking blades, organized around a central support, the rachis. This rachis is anchored in the skin by a cluster of muscles that the bird uses to tweak each feather's position. The wing is therefore a coordinated collection of smaller wings, giving the bird the supreme control that commands our admiration.

As the hawk moves through the forest, the feathers deflect air down, pushing the wing upward. Air also flows faster over the down-curved upper surface of the wing than it does over the concave underside. Fast-flowing air exerts less pressure, so the bird gets another lift. To land or change direction rapidly, she holds her wings at a sharp angle, breaking the smooth flow of air. The turbulence behind the wing acts as a brake, sucking the wing backward. The hawk's command of this stall is so refined that alighting motionless on a twig seems easy.

The hawk in the mandala was hunting. Sharp-shinned hawks feed mostly on small birds, such as winter wrens, and the hawk's wide, short wings let her slip between branches and accelerate powerfully as she pursues her prey. She uses her long tail to rudder through the tangled forest and snap upward, snatching flying birds from below with sicklelike talons. Any prey that escapes into tree holes or thickets is extracted with her gangly legs.

The hawk's design has one drawback. Rounded wings create turbulence at their blunt tips, spilling messy swirls of air. These swirls drag on the bird, making sustained flight more expensive for sharp-shinned hawks than it is for falcons and other angular-winged species. In addition, the wings of sharp-shinned hawks are not fanlike enough to allow them to soar like vultures. This is a bird of the forest, at home darting through pine and oak branches, and her design is ill suited to long flights. Sharp-shinned hawks cover long distances by alternating flapping with short glides, using a compromise between the continuous flapping of a falcon and the easy drifting of a vulture. This is tiresome work, and the hawks must stop to feed and rest along their way, unlike more accomplished long-distance flyers.

Sharp-shinned hawks in Tennessee do not migrate, but they are joined by sharp-shinned hawks retreating from winter farther north. This autumnal flow of southbound sharp-shinned hawks has dwindled in recent years. Scientists first suspected that pollution or habitat loss was causing the falling numbers of migrating hawks. But this is apparently not the case. Instead, more sharp-shinned hawks are choosing to stay in the frozen northern forests rather than head south for the winter. These lingering hawks survive by loitering around human settlements, making use of a remarkable new arrangement in the ecology of North America: the backyard bird feeder.

Our love of birds has created a new migration. This novelty is a west-to-east migration of plants, not a north-to-south migration of birds. The productivity of thousands of acres of former prairie land is shipped eastward, locked in millions of tons of sunflower seeds. These dense stores of energy are trickled from wooden boxes and glass tubes, adding a steady, stationary source of food to the otherwise unpredictably shifting winter food supply of songbirds in the eastern forest. Sharp-shinned hawks are therefore provided with a dependable meat locker, turning the forest into a home for the winter. Bird feeders not only augment the forest's larder but, more important, they gather songbirds into clusters that make convenient feeding stations for hawks.

The expression of our yearning for the beauty of birds sets off waves that circle outward, washing over prairies and forests, lapping onto the mandala. Fewer migrant hawks from the north make life a little easier for the hawk in the mandala. Winter becomes less dangerous for songbirds also, perhaps edging up winter wren populations. More abundant wrens may nudge down ant or spider populations, sending an eddy out into the plant community when the spring ephemeral flowers offer their seeds to be dispersed by ants, or into the fungus community when a dip in spider numbers increases fungus gnat populations.

We cannot move without vibrating the waters, sending into the world the consequences of our desires. The hawk embodies these spreading waves, and the marvel of its flight startles us into paying attention. Our embeddedness is given a magnificent, tangible form: here is our evolutionary kinship splayed out in the fanning wing; here is a solid, physical link to the north woods and the prairies; here is the brutality and elegance of the food web sailing across the forest.

November 21st—Twigs

The branches over the mandala are entirely bare. They fragment my view of the clear sky with a tracery of dark lines. Directly above me, a squirrel balances on impossibly thin twigs at the top of a maple tree. The squirrel's back feet grasp a twig while its forelegs and mouth reach out for clusters of seeds that have not yet fallen. Seed husks and small twigs rain in the animal's wake, pelting the ground with their hard falls. Whole seeds also drift down, spinning slowly in the breeze and landing several meters west of the mandala. This is the first time in several weeks that I have seen a squirrel in the maple tree. Lately the hickory's big, fatty nuts offered a better reward, but the nuts are gone and the squirrel has moved on to less favored foods.

One of the larger casualties of the squirrel's destructive foraging lies in front of me. The maple twig is half the length of my forearm, and its tip is branched into several clusters of empty seed stalks. At first, my eye passes over it lightly, unconsciously dismissing it. Then my glance backtracks, and detail explodes. Fungi have yet to blur the inscriptions on the twig's bark, so the story of this tiny piece of the canopy is writ clear.

The tan skin of the twig is scattered with creamy mouths, each oriented with lips opening parallel to the twig's length. These are lenticels, just visible to the naked eye, through which air flows to the cells below. As the twig matures to a branch, then a trunk, lenticels become less numerous and are hidden at the base of cracks in the bark. Younger twigs require a high density of lenticels to support their active, growing cells, just as a child's lungs are larger in proportion to her body than are those of an adult.

Larger swollen crescents rise from the surface where leaves formerly sprouted. Each leaf scar is topped by a small bud, or a circular indentation where the bud once grew. Twigs will grow from these buds, then most of the twigs will die within a year, a seemingly wasteful way to grow. After several years, only one or two twigs from hundreds remain, thickened into branches. This extravagance is a general theme of life's economy. Our nervous system also develops by ramifying into a complex web, then dying back to a simpler mature state. Social interactions do the same. The constant bickering among the members of a newly formed bird flock soon resolves itself into a simpler hierarchy where birds squabble only with their immediate superiors and subordinates.

Trees, nerves, and social networks are all systems that grow in unpredictable conditions. It is not possible for a maple seedling to know where the light will be strongest, or a nerve network to know what it will be called upon to learn, or a chick to know where it will fit in the social order. So, trees, nerves, and social hierarchies try dozens or hundreds of variants and pick out the best, molding themselves to their environments. Competition for light determines which twigs will live and die, and the varied architecture of trees grows out of the particularities of these hundreds of small events. A tree grown in the expansive light of an open field has a fan of branches that starts low on the trunk and gives the tree a wide, rounded profile. Here in the mandala, trees have few low branches and tight, cylindrical crowns, the result of crowding and competition for light. This process is analogous to evolution by natural selection, whereby a few winning characters are picked from the thousands of variants in each species. The process is already visible in the short length of twig in front of me. The older portion is bare, having already shed all its side branches, whereas the tip forks in a bush of curved matchsticks.

The smooth skin of the twig is interrupted by clusters of fine bracelets. These rings are the scars left by bud scales, the scooplike coverings that protect dormant buds through the winter. The tree's efforts to shield its growing tips etch a record of the passage of time, leaving a yearly ring of scars. The distance between the rings tells the vigor of that season's growth. Counting back from the tip: the maple twig grew an inch this year, an inch last year, and three inches in the two previous years. The oldest section was snapped by the squirrel's feet, but the remaining part shows six inches of growth. This twig has been slowing its growth for the past five years.

I turn my attention from the maple twig to the bud scales of the saplings in the mandala. Do they tell the same story as the maple twig? The green ash that sprouts knee-high from the mandala's center is topped by a magnificent bud, a swollen crown made of two large lobes flanked by two smaller teardrops. The bud scales that enclose this fat marvel are granular and the color of brown sugar. The marks from last year's scales sit just an inch below—not much growth this year. Last year was little better, but the year before that resulted in two inches of growth, and the four-year-old wood is very long, eight inches. Could the weather of the last two years have been unwelcoming in some way?

A maple sapling on the west side of the mandala shows the same pattern as the maple twig and the ash, although the difference among the years is less marked. The growth patterns of a maple and an ash two feet north break the pattern. Their twigs have grown more than ten inches for two years. These trees have been thriving, particularly on the branches that face east. Something more complicated than a uniform response to weather is affecting the trees' growth.

The variability in growth is partly caused by competition for light among the young trees. The collapsing growth rate of the ash in the mandala may result from the lush growth of the older ashes and maples that surround it. Four years ago, these older trees had not yet grown high enough to cast shade over the mandala's center. For the past three years they have progressively cast more shade, starving the ash.

The plants' growth is also affected by events beyond the local stem-to-stem race for light. There is a large hole in the forest canopy just to the east of the mandala. Two or three years ago an old shagbark hickory fell, dragging with it several smaller trees. I did not see this particular hickory crash down, but I have seen others fall. Tree falls start with the sound of rifle shots as wood snaps and the trunk fails. Loud hissing follows as thousands of leaves are dragged through the canopy, the sound rising as the tree accelerates. The impact of the trunk is like a huge bass drum, felt as much as heard. A wave of smell follows. Torn leaves give a sickly sweet odor that mingles with the bitter, wet smell of rent wood and bark. If the tree's roots were levered up by an unsnapped trunk, the ground is gouged and the root-ball stands up to six feet tall. The mess is impressive—smaller trees flattened, vines pulled from the canopy, twisted limbs everywhere. Once they are down, we can see what huge organisms trees are, like beached whales. A big tree fall can rip up a patch of forest the size of several houses, especially if it pulls other trees with it.

After the tree falls, light rushes in. Saplings that were not crushed or suffocated by fallen wood are bathed in light and grow fast. It has been a long wait. Although they are small and look young, some of these saplings may be tens or hundreds of years old. They grew slowly in the shade, died back to the roots every few years, then resprouted to creep up again, biding their time until the gap opened and released them.

The quality of light also changes under a break in the canopy. Leaves absorb some wavelengths of light better than others. In particular, they absorb red light, but "far red" light passes through them. Far red, or infrared, is invisible to humans because its wavelength is too long for the receptors in our eyes. But plants can "see" both red and far red light. Growing twigs use the relative proportions of the two wavelengths of light to tell where they are in relation to other plants.

Under the canopy and in crowded conditions, far red light dominates because leaves of competing plants absorb most of the red light. But under open skies the proportion of red light surges. Twigs respond by changing their architecture, spreading branches wide and stretching their tips into the light.

Trees accomplish "color vision" through a chemical in their leaves. This molecule, called phytochrome, can exist in two different forms. The switch between forms is activated by light: red causes the molecule to flip into the "on" position; far red causes it to flip "off." Plants use the two forms to assess the ratio of red and far red light in their environment. In the reddish light of the canopy gap, phytochromes in the "on" position dominate, causing trees to grow bushy branches toward the gap. In the forest shade, far red dominates, and trees grow upward with spindly trunks and few side branches. The whole plant is permeated with phytochromes, so trees act like large eyes, sensing color over their entire bodies. Ralph Waldo Emerson, who claimed to be a transparent eyeball, open to the woods, would perhaps appreciate the trees' superior abilities in this regard.

The vegetation directly under the gap is unmistakably transformed by the infusion of light, but the canopy's rupture also bleeds sunshine into the surrounding forest, even to the mandala, which is firmly nestled under an umbrella of maple and hickory. Saplings are growing faster on the eastern side, and branches that face this direction are more vigorous than their westerly peers. The mountain slope here has a northeastern tilt, so the light gap reinforces a preexisting bias.

The ground-hugging layer of herbaceous plants is also affected by the gap. Leafcups are absent from the western half of the mandala and grade upward from stunted plants at the mandala's center to vigorous ankle-high individuals at the eastern edge. These plants are adapted to growing in forest gaps and reach knee height in the center of the opening. The tallest plants will reach my shoulders next year when they flower after completing their second and final year of growth. The other herbaceous plants, *Hepatica* and cicely, show little sign of leaping at the light and seemingly grow just as well in the shady western half of the mandala as in the east. This surface uniformity may hide more subtle effects, because these plants respond to higher levels of light not by growing taller but by setting more seed or sending out more rhizomes.

Within five years, the gap will be choked with saplings racing for the canopy. Mature trees on the edges will reach into the gap, grabbing light from above the saplings. In ten years, one or two of the saplings will have won and the dozens of losers will be dying. This struggle is short in comparison to the centuries that mature trees live once they reach the canopy, but intense competition among young trees has a large effect on the forest's composition. In the diverse forests of Tennessee, no one species consistently wins the sprint for the canopy, a reflection of the variegated soil and temperate climate.

The fallen hickory tree and the broken twig are two points on a broad continuum of disturbance in the canopy. On one end of this continuum are massive disruptions such as hurricanes, which visit rarely, every hundred years at most in this part of Tennessee. At the other end of the spectrum are the tiny holes left in the canopy's scaffolding by the squirrel's damaging feet. These holes are short-lived and small-scale, creating the sunflecks that boost the growth of ephemerals and low-growing saplings. Wood rot and winter ice storms also create small holes in the canopy. I hear a large branch crashing down every few hours, particularly in winter. Disturbance on an intermediate scale is also common, with windstorms being by far the most regular source.

Storms in the forest have a more primal character than storms playing over tamed urban land. A vigorous cloudburst is exhilarating, a burst of sensory delight with its leafy smells, gray light, and chill. But a full tree-ripping storm pushes the senses beyond exhilaration or thrill, into fear. As rain patter turns to squall, the canopy heaves under the pressure of the wind. Tree trunks saw to and fro, flexing beyond what looks possible, then slashing back. All my senses wake, my eyes dart. Then the floor bucks. As trees oscillate, they yank on roots and lift the ground. As if I'm walking on the deck of a yawing boat, my feet stumble. The storm confuses—my eyes are blurred by streaming rain, ears jammed with the roar of wind in leaves; the ground lurches underfoot. This confusion focuses into an urge to run, but unless rocks or other shelter is nearby, running goes nowhere safe. Periodically a severed tree limb crashes through the branches. The imagination takes off, and every snap becomes the sound of falling timber. In these storms I'll either scamper to shelter, if it is available, or huddle against a sturdy-looking trunk, feeling its weight heave against my back. What I fear most is the fall of a full-sized tree, but the fear has no outlet, so I sit wide-eyed until the storm eases. At the storm's peak, there is strange comfort in my powerlessness. Nothing I do can influence the lashing world that I'm caught in, so surrender follows, and with it comes a curious state: mental clarity wrapped in an electrified body.

Violent storms hit the mountainside dozens of times each year. But they are short-lived, and their physical damage tends to be focused on small areas—a patch of older maples here, a loose-rooted giant buckeye there. The forest is pockmarked with the gaps formed by these falls. For some species, such as the sugar maples, a canopy opening offers an accelerated route to the top. But maples are shade tolerant, so their growth does not require canopy openings. For other species, however, gaps are the only hope. Tuliptrees and, to a lesser extent, oaks, hickories, and walnuts need bright light to grow, so the persistence of these species depends on the irregular patchwork of disturbance across the forest. The tuliptree seeds that landed in the shady part of the mandala have little hope of germinating and surviving their first year. Those that landed twenty feet east of here will slake their greedy thirst for sunlight and vie to be the one-in-a-million seed that fulfills its potential and reaches the canopy.

The canopy's renewal depends paradoxically on its being split open to let light reach the ground. Any change in the dynamics of gaps will therefore affect the viability of the forest. This makes me particularly concerned about the spindly tree growing at the back of the gap next to the mandala. This tree has grown several feet since the spring, thrusting its two-foot-wide, heart-shaped leaves into the opening. Fast-growing alien species such as this *Paulownia tomentosa*, or Princess Tree, are spreading through the eastern forests, taking over the forest by invading light gaps and outgrowing native species. *Paulownia*, and its invading partner, *Ailanthus altissima*, the Tree of Heaven, produce thousands of wind-dispersed seeds and thereby spread rapidly. They especially favor roadside edges and logged forests, but like most pioneers will readily invade openings after smaller-scale disturbances.

Fast-growing invaders are particularly harmful to the regeneration of native trees that require full sunlight to grow: oaks, hickories, walnuts, tuliptrees. When *Paulownia* and *Ailanthus* sprout in a gap, they smother the slower-growing natives. In forests that are heavily disturbed by fire, logging, or housing development, nonnative trees can quickly erode native tree diversity.

The study of twigs seems esoteric. But this impression is dangerously wrong. Counting back through bud scars, tallying yearly growth, I not only see the struggle among native and foreign trees, I read the ledger of the world's atmosphere. Each twig yearly adds a few inches, and these inches, combined across the forest, create one of the world's biggest stores of carbon.

When we count all new growth—twigs, leaves, thickening trunks, extended roots—the mandala likely took ten or twenty kilograms of carbon from the air this year, a pile of sticks about as big as a small car. Summed over the world's surface, forests contain about twice as much carbon as the atmosphere, over a thousand million million tons. This vast store is our buffer against calamity. Without forests, much of the carbon would be in the air as carbon dioxide gas, baking us in an awful greenhouse.

As we've burned oil and coal, we've returned long-buried carbon stores to the atmosphere. But forests have saved us from the full brunt of the resulting climate change. Half our burned carbon has been absorbed by forests and the oceans. Lately, this buffering effect of forests has diminished—there is a limit to the rate at which trees can soak extra carbon from the atmosphere, especially as we accelerate our burning of fossil fuel. Nonetheless, forests continue to shield us from the more dire consequences of our profligacy. The study of twigs and bud scars is therefore the study of our future well-being.

December 3rd—Litter

I lie facedown at the edge of the mandala, readying myself for a dive under the surface of the leaf litter. The red oak leaf below my nose is crisp, protected from fungi and bacteria by the drying sun and wind. Like the other leaves on the litter's surface, this oak leaf will remain intact for nearly a year, finally crumbling in next summer's rains. These surface leaves form a crust that both hides and makes possible the drama below. Protected under the shield of superficial leaves, the rest of autumn's castoffs are pulverized in the wet, dark world of the litter. Yearly, the ground heaves like a breathing belly, swelling in a rapid inhalation in October, then sinking as the life force is suffused into the forest's body.

Below the red oak leaf, other leaves are moist and matted. I tease away a wet sandwich of three maple and hickory leaves. Waves of odor roll out of the opening: first, the sharp, musty smell of decomposition, and then the rounded, pleasant odor of fresh mushrooms. The smells are edged with a richer, earthy background, the mark of healthy soil. These sensations are the closest I can come to "seeing" the microbial community in the soil. The light receptors and lenses in my eyes are too large to resolve the photons bouncing off bacteria, protozoa, and many fungi, but my nose can detect molecules that waft out of the microscopic world, giving me a peek through my blindness.

A peek is about all that anyone is given. Of the billion microbes that live in the half handful of soil that I have exposed, only one percent can be cultivated and studied in the lab. The interdependencies among the other ninety-nine percent are so tight, and our ignorance about how to mimic or replicate these bonds is so deep, that the microbes die if isolated from the whole. The soil's microbial community is therefore a grand mystery, with most of its inhabitants living unnamed and unknown to humanity.

As we chisel away at the edges of this mystery, jewels fall out of the eroding block of ignorance. The earthy smell that embraces my nose comes from one of the brightest jewels, the actinomycetes, strange semicolonial bacteria from which soil biologists have extracted many of our most successful antibiotics. Like the healing chemicals in foxglove, willow, and spirea, the actinomycetes use these molecules in their struggle with other species, secreting antibiotics to subdue or kill their competitors or enemies. We turn this struggle to our advantage through medicinal mycology.

Antibiotic production is a small part of the huge and varied role of actinomycetes in the soil's ecology. There is as much diversity within the feeding habits of this group of bacteria as exists within all the animal kingdom. Some actinomycetes live as parasites in animals; others cling to plant roots, nibbling on them while fighting off more damaging bacteria and fungi. Some of these root dwellers may turn against their hosts and kill the plant by belowground assassination. Actinomycetes also coat the dead bodies of larger creatures, breaking them down into humus, the dark miracle ingredient of productive soils. Actinomycetes are everywhere but seldom enter our consciousness. Yet we seem to have an intuitive understanding of their importance. Our brains are wired to appreciate their distinctive "earthy" smell and to recognize the aroma as the sign of good health. Soil that has been sterilized, or that is too wet or dry for most actinomycetes, smells bitter and unfriendly. Perhaps our long evolutionary history as hunter-gatherers and agrarians has taught our nasal passages to recognize productive land, giving us a subconscious tie to the soil microbes that define the human ecological niche.

The other members of the microbial community are harder to pin down in the complex smell that rises from the earth's abdomen. Fungal spores contributed to the acrid mustiness; bacterial decomposers released sweet aromas from the remains of dead leaves. Tiny wafts of methane rise up from sodden patches where anaerobic microbes hide. Many other microbes live beyond the reach of my nose. Bacteria grab nitrogen from the air and pass it into the biological economy. Others take nitrogen from dead creatures and send it back into the air. Protists graze on fungi and bacteria that encrust decaying leaves. This secret microbial world has existed for a billion or more years. The bacteria, in particular, perform biochemical tricks that have fed them since the earliest years of life, three billion years ago. The smell in my nose therefore comes from a hidden world that is broad and deep, complex and ancient.

Microbes may be invisible, but my window into the soil offers plenty else to see. Lightning-white fungal strands crackle over black leaves. Pink hemipteran bugs dance around orange spiders. A ghostly white springtail moves over the dark crumbs of last year's decayed leaves. Everything lives in miniature. A buried maple seed towers over the animals like a mansion dwarfing its owner. The largest living creature is a rootlet, one tiny part of a plant, perhaps a sapling or tree. It is barely thicker than a pin, but it dominates the small hole that I have bored into the litter.

The rootlet is a smooth, creamy cable sprouting a haze of hairs that radiates out into the soil's matrix. Each of these hairs is a delicate extension of the root's surface, a tentacle stretched out from a plant cell. The hairs creep around sand particles, sliding into the films of water that cling to the soil. By greatly increasing the surface area of the root, these hairs allow the plant to harvest water and nutrients that would otherwise be unavailable. So critical are these root hairs that if their intricate hold on the soil is broken by being uprooted or transplanted, the plant will wilt and die unless it receives extra watering from a gardener.

Root hairs draw water and dissolved nutrients from the soil, sending them upward to slake the leaves' thirst and to supply the minerals that plants need for their building projects. The energy for this skyward motion comes mostly from the sun's evaporative power, transmitted downward through columns of water in the xylem. But the root hairs are not just passive pipe ends sucking at the soil like a pump in a well. Their relationship with the physics and biology of the soil is reciprocal.

The simplest of the roots' gifts to the soil are hydrogen ions, pumped out by the root hairs to help loosen nutrients that are bound to clay particles. Each fleck of clay has a negative charge, and so minerals with positive charges such as calcium or magnesium stick to the clay's surface. This attraction helps the soil hold on to its minerals, stopping them from washing away in the rain, but the bond also prevents plants from acquiring them in the flow of water into the root. The root hairs' answer is to soak the clay particles in positively charged hydrogen ions. These dislodge some of the attached mineral ions from the surface of the clay. The released minerals float in water films that surround the clay and are swept into the root hairs by the water's flow. The most useful of these minerals are easily dislodged, so the root hairs need release only a small amount of hydrogen ions to receive their reward. More vigorous applications of hydrogen ions, such as those that come with acid rain, release the more toxic elements such as aluminum.

Roots also supply the soil with large amounts of organic matter. Unlike the deposition of leaves from above, most of the roots' donations are actively secreted, not cast off as waste products. Dead roots certainly enrich the soil, but death's contribution is dwarfed by the tangle of sugars, fats, and proteins that living roots infuse into the soil around them. This gelatinous sheath of food around the root creates a buzz of biological activity, particularly near the root hairs. As in a sandwich shop at lunchtime, much of the soil's life is crowded into the narrow root zone or rhizosphere. Here, microbial densities are a hundred times higher than in the rest of the soil; protists crowd around, feeding on microbes; nematodes and microscopic insects push through the crowd; fungi spread their tendrils into the living soup.

The ecology of the rhizosphere is mostly a mystery, made difficult to study by its paper-thin delicacy. Plants obviously stimulate life in the soil, but what do they receive in return? The explosion of biological diversity in the rhizosphere may protect roots from disease, just as a diverse forest is less likely than a bare field to be overrun with weeds. But this is speculation. We are explorers standing at the edge of a dark jungle, peering at the strange shapes in the soil's interior, naming a handful of the most obvious novelties but understanding little.

Despite the gloom, one relationship in the jungle of the rhizosphere is so important that even the most hasty explorers trip over its vines, then look up, astounded. The plants' partners in this surprising relationship are visible in the window that I have created into the leaf litter. Fungal threads cover most of the soil like a subterranean spiderweb. Some are dusky gray and spread out seemingly at random, coating whatever lies in their path. Others grow their white strands in waving lines, diverging then reuniting like rivers in a delta. Each fungal thread, or hypha, is ten times finer than a root hair. Because hyphae are so thin, they can squeeze between microscopic soil particles and penetrate the ground much more effectively than can clumsy roots. A thimbleful of soil may contain a few inches of root hairs but a hundred feet of hyphae, spooled around every fleck of sand or silt. Many of these fungi work alone, digesting the decaying remains of leaves and other dead creatures. Some, however, work their way into the rhizosphere and begin a conversation with the root. This conversation is the start of an ancient and vital relationship.

The fungus and the root greet each other with chemical signals and, if the salutation goes smoothly, the fungus extends its hyphae in readiness for an embrace. In some cases, the plant responds by growing tiny rootlets for the fungi to colonize. In others, the plant allows the fungus to penetrate the root's cell walls and spread the hyphae into the interior of the cells. Once inside, the hyphae divide into fingers, forming a miniature rootlike network within the cells of the root. This arrangement looks pathological. I would be a sick man if my cells were infested with fungi in this way. But the ability of hyphae to penetrate plant cells is put to healthy use in this marriage with roots. The plant supplies the fungus with sugars and other complex molecules; the fungus reciprocates with a flow of minerals, particularly phosphates. This union builds on the strengths of the two kingdoms: plants can create sugars from air and sunlight; fungi can mine minerals from the soil's tiny crevices.

The fungus-root, or mycorrhizal, relationship was first discovered as a spin-off from the King of Prussia's attempts to cultivate truffles. His biologist failed to domesticate the valuable fungus, but he discovered that the underground fungal network that produces truffles is connected to tree roots. He later showed that these fungi were not parasites, as he first suspected, but acted as "wet nurses" passing nutrients to trees and increasing their rate of growth.

As botanists and mycologists worked their way across the plant kingdom, peering at root samples through microscopes, they found that nearly all plants have mycorrhizal fungi wrapped in or around their roots. Many plants cannot live without their fungal partners. Others can grow alone but are stunted and weak if they cannot meld their roots with a fungus. In most plants, fungi are the main absorbing surface in the soil; roots are just the connections to this network. A plant is therefore a paragon of cooperation: photosynthesis is made possible by ancient bacteria embedded in its leaves, respiration is likewise powered by internal helpers, and roots serve as connectors to an underground network of beneficial fungi.

Recent experiments show that mycorrhizae take this relationship yet further. By feeding plants radioactive atoms, plant physiologists have traced the flow of matter in the forest ecosystem and found that fungi act as conduits among plants. Mycorrhizae are promiscuous in their embrace of plant roots. Seemingly independent plants are physically connected by their subterranean fungal lovers. The carbon taken out of the atmosphere and turned into sugar by the maple tree above the mandala may find itself transported to the tree's roots and donated to a fungus. The fungus will then either use the sugar for itself or pass the sugar to the hickory, or to another maple, or to the spicebush. Individuality is therefore an illusion in most plant communities.

Ecological science has yet to fully digest the discovery of the below-ground network. We still think of the forest as being ruled by relentless competition for light and nutrients. How does the mycorrhizal sharing of resources change the aboveground struggle? Surely the race for light is no illusion? Could some plants be parasitizing others, using the fungi as friendly con men, or do the fungi mitigate and smooth out differences among plants?

Whatever the answers to these questions, it is clear that the old "red in tooth and claw" view of the natural economy has to be updated. We need a new metaphor for the forest, one that helps us visualize plants both sharing and competing. Perhaps the world of human ideas is the closest parallel: thinkers are engaged in a personal struggle for wisdom, and sometimes fame, but they do so by feeding from a pool of shared resources that they enrich by their own work, thus propelling their intellectual "competitors" onward. Our minds are like trees—they are stunted if grown without the nourishing fungus of culture.

The partnership between fungi and plants that undergirds the mandala is an old marriage, dating to the plants' first hesitant steps onto land. The earliest terrestrial plants were sprawling strands that had no roots, nor any stems or true leaves. They did, however, have mycorrhizal fungi penetrating their cells, helping to ease the plants into their new world. Evidence of this partnership is etched into fine-grained fossils of the plant pioneers. These fossils have rewritten the history of plants. The roots that we thought were one of the earliest and most fundamental parts of the land plant body turn out to be an evolutionary afterthought. Fungi were the plants' first subterranean foragers; roots may have developed to seek out and embrace fungi, not to find and absorb nutrients directly from the soil.

Thus cooperation gains another jewel in its evolutionary crown.

Most of the major transitions in life's history were accomplished through joint ventures such as the union between plant and fungus. Not only are the cells of all large creatures inhabited by symbiotic bacteria, but the habitats they live in are made by or modified by symbiotic relationships. Land plants, lichens, and coral reefs are all products of symbiosis. Strip the world of these three and you have stripped it virtually bare—the mandala would be transformed to a pile of rocks clothed in bacterial fuzz. Our own history mirrors this pattern: the agrarian revolution that unleashed humanity's boom was created by entering into mutual dependence with wheat, corn, and rice, and by conjoining our fate with that of horses, goats, and cattle.

Evolution's engine is fired by genetic self-interest, but this manifests itself in cooperative action as well as solo selfishness. The natural economy has as many trade unions as robber barons, as much solidarity as individualistic entrepreneurship.

My peephole into the soil gave me a glimpse into some new ways of thinking about evolution and ecology. Or are they so new? Perhaps soil scientists are rediscovering and extending what our culture already knows and has embedded into our language. The more we learn about the life of the soil, the more apt our language's symbols become: "roots," "groundedness." These words reflect not only a physical connection to place but reciprocity with the environment, mutual dependence with other members of the community, and the positive effects of roots on the rest of their home. All these relationships are embedded in a history so deep that individuality has started to dissolve and uprootedness is impossible.

December 6th—Underground Bestiary

Our everyday experience of the animal kingdom is dominated by two groups of animals: the vertebrates and the insects. These two branches on the tree of life occupy most of our culture's zoological field of vision, yet they represent just a fraction of the structural diversity of animals. Biologists divide the animal kingdom into thirty-five groups or phyla, each one defined by a distinctive body plan. The vertebrates and insects represent two subphyla among the thirty-five.

Why have the birds and the bees captured our imagination, leaving the nematodes, flatworms, and the rest of the world's bestiary in a dusty back room of our consciousness? The simple answer is that we don't run into nematodes very often. Or, we think we don't. A deeper answer seeks to explain why the larger part of animal diversity is hidden from us. We get out and about enough, so why do we not run into our neighbors?

Unfortunately for the richness of our experience, we live in a strange and extreme corner of the world's available habitat. The animals that we encounter are those few that also inhabit this unusual niche.

The first cause of our estrangement is our size. We are tens of thousands of times larger than most living creatures, therefore our senses are too dull to detect the citizens of Lilliput that crawl around and over us. Bacteria, protists, mites, and nematodes make their homes on the mountains of our bodies, hidden from us by the dislocation of scale.

We live in the empiricist's nightmare: there is a reality far beyond our perception. Our senses have failed us for millennia. Only when we mastered glass and were able to produce clear, polished lenses were we able to gaze through a microscope and finally realize the enormity of our former ignorance.

Our living on land further distances us from the rest of the animal kingdom, augmenting the handicap of gigantism. Nine-tenths of the animal kingdom's main branches are found in water—in the sea, in freshwater streams and lakes, in watery crevices within the soil, or in the moist interiors of other animals. The desiccated exceptions include the terrestrial arthropods (mostly insects) and the minority of vertebrates that have hauled themselves onto land (most vertebrate species are fish, so terrestrial life is unusual even for a vertebrate). Evolution has plucked us out of our wet burrows, leaving our kin behind. Our world is therefore populated by extremists, giving us a distorted view of life's true diversity.

My first dive into the soil helped me escape my strange ecological hermitage and gave me a taste of the treasury that lives below the surface. My thirst is whetted, so I dip down again. At three spots around the edge of the mandala I peel away a small clump of leaves, create a small hole in the litter, peer down with my hand lens, then replace the leaves. The contrast with the aboveground world is striking. Aboveground, apart from a passing titmouse, I seem to be alone in the forest. Yet animals abound an inch below the surface of the litter.

The largest animal in my foray is a salamander curled in the cup of a fallen oak leaf. The salamander would fit on my thumbnail but is hundreds of times larger than all the other animals that I encounter. This salamander is a crocodile among minnows, watched by a nearsighted whale.

As I peer closely through the hand lens, behind the salamander I see flickering movement and minute undulations on fungal strands and dead leaves. I strain my eyes until they ache, but I cannot identify the tiny animals that create these movements. I have reached a perceptual wall. Fortunately there is plenty to see on this side of the wall. The most common creatures are the springtails, or collembolans. If the mandala is typical of most terrestrial ecosystems, there may be up to a hundred thousand springtails within its borders. It is therefore not surprising that I find at least one of these tiny animals every time I lift a leaf. To the naked eye they are undefined flecks, but under the hand lens I can make out six stumpy legs protruding from the barrel-like body. All the individuals that I examine are doughy white and wet, with no eyes. These animated jellybeans are members of the onychiurid family of springtails. Their lack of pigment and blindness reflects the subterranean specialization of the group; unlike other springtails, these animals never wander aboveground. The onychiurids have lost the jumping organ, the furca, that gives springtails their name. Presumably a powerful belly-mounted catapult is of little use to an animal that spends its life in the soil's crevices. Instead of springing away from predators, the onychiurids deter them by releasing noxious chemicals from glands on their skin. These chemicals ward off predatory mites and other common soil carnivores but are presumably less effective against the larger but less frequently encountered beaks of wrens and turkeys.

One hundred thousand springtails produce a lot of mini dung pats. The mandala contains a million springtail pellets, each one a miniature package of composted fungus or plant. The spores of bacteria and fungi pass through the gut undigested, so springtails act both as dispersers of the microbial community and as the soil's champion compost makers. Springtails also exert an important effect at the other end of their digestive tracts. Although the details of this relationship are still unclear, springtails seem to enhance the mycorrhizal association between fungi and plant roots. They graze on fungal strands and thereby stimulate some fungi and suppress others. Springtails are like cows on a pasture, regulating the growth of their food by continuous clipping and by fertilizing the ground with their droppings.

The springtails' central position in the life of the soil is, unfortunately, not reflected in their taxonomy. They have six legs, but their strange mouthparts (housed in a reversible pouch in the head) and distinctive DNA indicate that they are a sister group to the insects. Because they fall between the insects and the other invertebrates, the springtails are claimed by few biologists, and their lives are poorly known. But they are the evolutionary soil from which came the insect inhabitants of our aboveground world.

Springtails are the most numerous animals in my samples, but because springtail bodies are small, they account for less than five percent of the total weight of animals in forest soil. Relative to their ecological importance, springtails are also a species-poor group of animals. Six thousand springtail species grace the planet, compared to a million insect species (and over one hundred thousand species of fly). As I move around the mandala, I therefore encounter many springtails, all seemingly of the same type. The other animals that I find are different in each sample, reflecting their higher taxonomic diversity.

After the springtails, the next most abundant visible animals are other arthropods: spiders, gnats, and millipedes. The armored body plan of the arthropods has been modified by evolution into an engineer's fantasy world of designs. Armor has been flattened into the wings of flies and sharpened into spider fangs. Jointed legs have been turned into silk-spinning tweezers, mushroom-munching mouthparts, and all-terrain climbing boots. No other group of animals rivals the arthropods in the diversity of body forms, yet all these forms are built on the same ground plan: a segmented outer case that is molted periodically to allow new growth.

The arthropod body plan is well represented in the mandala, but it is not the only one. Tiny snails graze among the dead leaves in the mandala's soil. Some of these are juvenile versions of the larger snails that graze on the surface of the mandala, but other species spend their entire lives in the wet embrace of the litter. The snail's shell is excellent armor but is simpler and less versatile than the arthropod's all-encasing jointed suit. Because snails don't molt, they cannot fully enclose their body within the shell. Snails are therefore vulnerable to attack through their shells' openings. Many of the mandala's snails have reduced this risk by partly barring their shell's mouth with toothlike extensions to the shell's lip. Some of these extensions are so well developed that there is little room for the snail's body to squeeze between them as the fleshy part of the animal pokes out of the shell to feed.

Snails owe their success to the clever ways they use their tongues. They are the world's most successful lickers—few surfaces on the planet escape their attention. The tongue, the radula, is a toothy strap that the snail pokes out, then scrapes back, rasping whatever is below. As the radula moves back into the mouth it passes over a stiff lower lip, causing the radula to fold back and the teeth to bristle up. Each tooth is a bulldozer blade that gouges into the surface below, shoveling food into the mouth. This cross between a conveyor belt and carpenter's plane is the key that unlocks the world for snails. We look at a boulder and see bare rock; snails experience a film of butter and jelly spread over the boulder's surface.

As I continue my subterranean dive, I find yet another body form, the "worm." Some of these worms appear familiar—the segmented earthworms and diminutive relatives of earthworms, the enchytraeids, or potworms. But my attention rests on these familiar figures for a few seconds, then is drawn to another, stranger worm that rests on the torn edge of a leaf. This animal is visible only through my hand lens and sits in the film of water that covers the leaf. As I watch, the worm rears up and thrashes in the air, then falls back into the water. The lurching motion identifies the worm as a nematode. Unlike the earthworms and potworms, this animal has no body segments, and its head and tail taper to points. There may be a billion nematodes in the mandala, most of them so small that only a strong microscope can reveal them. Some are parasitic, some are free-ranging predators, others graze on plants and fungi. Only the arthropods have more diverse feeding styles and ecological roles. Yet because nematodes are so small and water-loving, they live in scientific obscurity. The few people who do study these worms boast that if all matter were removed from the universe, leaving only nematodes, the planet's shape would remain, a haze of worms. The forms of animals, plants, and fungi would still be discernible in the creamy fog and, because nematodes are so specialized, the original inhabitants of these forms would still be identifiable. Tell me what worms you have and I'll tell you who you are.

My foray into the upper surface of the mandala's soil has revealed more diversity of animal body designs than could be found in all of a zoo's displays. Multitudes crawl, squirm, and writhe below my feet. But I seem to be alone up here in the air above the mandala. The soil's warmth and moistness help make possible the zoological extravagance, but these benign conditions would be for naught if the soil were not well provisioned. Death is the soil's main supplier of food. All terrestrial animals, leaves, dust particles, droppings, tree trunks, mushroom caps, are destined for the soil. We are all destined to pass through the dark underworld, feeding other creatures as we go. The human economy has no counterpart to the soil's all-embracing monopoly. Some segments of our economy have more power than others, but nowhere does any one industry get to process and profit from the work of *all* the others. Banks come close, but the cash economy bypasses them. In nature, however, there is no escape from Isaiah's prophecy; "their root shall moulder away, and their shoots vanish like dust." Decomposers and their business partners fill the soil with their lively and varied activities. The seeming dominance of the aboveground world is therefore an illusion. At least half the world's activity is belowground.

In the end, it is not just the diversity of the bestiary that our size and dryness hides from us but the true nature of life's physiology. We are bulky ornaments on life's skin, riding the surface, only dimly aware of the microscopic multitudes that make up the rest of the body. Peering below the mandala's surface is like resting lightly on the body's skin, feeling the pulse move.

December 26th—Treetops

It is midday and the sky is clear, but no sun shines on the mandala. The slope here is tilted northeast, away from the low sun, and the bluff uphill blocks direct sunlight. Slanting rays clear the bluff and illuminate treetops, creating a divide between light and shadow that cuts trees at twelve feet from the ground. This divide will sink day by day until the sun is high enough, in February, to kiss the ground again after a long absence.

Four gray squirrels loaf in the bright upper branches of a dead shagbark hickory tree fifty meters down the slope. I watch them for an hour, and mostly they loll in the sun, limbs sprawled. They seem companionable, sporadically nibbling the fur on one another's hind legs or tails. Occasionally one will break from sunbathing and chew the fungus-encrusted dead branches, then return to sit silently with the other squirrels.

This scene of sciurid tranquillity makes me unaccountably delighted. Perhaps I so often see and hear squabbling among the squirrels that today's ease seems particularly sweet. But something more is behind my delight; I feel freed from some burden carried by my overtrained mind. Wild animals enjoying one another and taking pleasure in their world is so immediate and so real, yet this reality is utterly absent from textbooks and academic papers about animals and ecology. There is a truth revealed here, absurd in its simplicity.

This insight is not that science is wrong or bad. On the contrary:

science, done well, deepens our intimacy with the world. But there is a danger in an exclusively scientific way of thinking. The forest is turned into a diagram; animals become mere mechanisms; nature's workings become clever graphs. Today's conviviality of squirrels seems a refutation of such narrowness. Nature is not a machine. These animals feel. They are alive; they are our cousins, with the shared experience that kinship implies.

And they appear to enjoy the sun, a phenomenon that occurs nowhere in the curriculum of modern biology.

Sadly, modern science is too often unable or unwilling to visualize or feel what others experience. Certainly science's "objective" gambit can be helpful in understanding parts of nature and in freeing us from some cultural preconceptions. Our modern scientific taste for dispassion when analyzing animal behavior formed in reaction to the Victorian naturalists and their predecessors who saw all nature as an allegory confirming their cultural values. But a gambit is just an opening move, not a coherent vision of the whole game. Science's objectivity sheds some assumptions but takes on others that, dressed up in academic rigor, can produce hubris and callousness about the world. The danger comes when we confuse the limited scope of our scientific methods with the true scope of the world. It may be useful or expedient to describe nature as a flow diagram or an animal as a machine, but such utility should not be confused with a confirmation that our limiting assumptions reflect the shape of the world.

Not coincidentally, the hubris of narrowly applied science serves the needs of the industrial economy. Machines are bought, sold, and discarded; joyful cousins are not. Two days ago, on Christmas Eve, the U.S. Forest Service opened to commercial logging three hundred thousand acres of old growth in the Tongass National Forest, more than a billion square-meter mandalas. Arrows moved on a flowchart, graphs of quantified timber shifted. Modern forest science integrated seamlessly with global commodity markets—language and values needed no translation.

Scientific models and metaphors of machines are helpful but limited. They cannot tell us all that we need to know. What lies beyond the theories we impose on nature? This year I have tried to put down scientific tools and to listen: to come to nature without a hypothesis, without a scheme for data extraction, without a lesson plan to convey answers to students, without machines and probes. I have glimpsed how rich science is but simultaneously how limited in scope and in spirit. It is unfortunate that the practice of listening generally has no place in the formal training of scientists. In this absence science needlessly fails. We are poorer for this, and possibly more hurtful. What Christmas Eve gifts might a listening culture give its forests?

What was the insight that brushed past me as the squirrels basked? It was not to turn away from science. My experience of animals is richer for knowing their stories, and science is a powerful way to deepen this understanding. Rather, I realized that all stories are partly wrapped in fiction—the fiction of simplifying assumptions, of cultural myopia and of storytellers' pride. I learned to revel in the stories but not to mistake them for the bright, ineffable nature of the world.

December 31st—Watching

The weak sun of late afternoon shines on the westward facing slope of the other side the valley. The red-tinged light reflects from the bark of massed trees, giving the forest a purple-gray glow. As the sun falls, a line of shadow swings up the slope, extinguishing the warm reflection, turning the forest to dusky brown. As the sun drops lower, its rays angle into the sky, over the mountain. Crimson shades into haze on the horizon, and the sky's blue fades, first to watery mauve, then to gray.

Ten days ago, on the day of the winter solstice, I watched this same swing of sunlight. The rising border between dark and light on the opposite forest slope drew all my attention, its climb up the mountain building to the moment when the shadow would crest and the bright sunlight would blink out. At the very instant that the line of shadow hit the horizon, coyotes hidden on the forest slope just to my east broke into howls. They yipped and wailed for half a minute, then fell silent. The timing of their chorus seemed too precise to be a coincidence, coming as it did at the moment that the sun slipped off the slope. We may both, coyote and human, have watched the bright spectacle on the mountainside and been stirred by the sight of the sun's disappearance. Coyotes' howling behavior is known to be sensitive to both daylight and to the moon's phase, so it is not unreasonable to suppose that these animals might sometimes wail to the setting sun.

This evening, the coyotes are either hushed or absent, and I watch the changing light without their accompaniment. The forest is not silent, however. Birds are particularly vocal, perhaps enlivened by the day's temperature, which climbed well above freezing. Now, the wrens and woodpeckers chatter as they go to roost, chipping and scolding as darkness thickens. When the sun has fallen well below the horizon and the fussing birds have quieted, a barred owl yelps from high in a tree just down the slope. The owl repeats its strangled barks a dozen or more times, perhaps calling to a mate in this winter season of owl courtship.

After the owl falls silent, the forest enters a deeper quiet than I remember experiencing here. No birds or insects call. The wind is still. The sounds of human activity, distant aircraft or roads, fall away. The very soft murmur of a stream to the east is the only detectable sound. Ten minutes pass in this peculiar calm. Then the wind quickens, drawing a hiss from the treetops. A high airplane rumbles, and muffled hammering echoes up the valley from a distant farm. Each sound is made vivid by the surrounding silence.

The horizon bleeds away its color and luminosity, falling into deep blue. The fat-bellied moon, three-quarters full, shines low in the sky. My eyes lose their power as the forest turns to shadow.

The stars slowly kindle from the sky's darkness. The day's energy ebbs, leaving me at ease. Suddenly—stab!—a blade pierces me. Fear. The coyotes rip open the calm. They are close, much closer than ever before. Their crazed howling comes from just a few meters away. The sound crescendos in squeals and whistles, overlain on deep-throated barking. My mind transforms immediately. The blade focuses all thought: wild dogs will tear you apart. Hell, they are *loud*.

All this in just a few seconds. Then my conscious mind reasserts itself, and before the chorus is over I have dislodged the blade. There is no chance that these coyotes will bother me. Rather, I'm lucky that they did not pick up my scent, or they would not have come so near. My fear passes quickly. But, for a moment, my body remembered ancient lessons. The focused memory of hundreds of millions of years of hunted life exploded in my head with utter clarity.

The coyotes' chorus carries for miles down the valley, setting off farm dogs in distant barns and fields. Dog minds have also been shaped by years of selection, encouraged by our agrarian ancestors to bark incessantly after hearing the howls of wild relatives. No coyote or wolf would dare penetrate a cacophony of farm dogs, and this fear gives vulnerable livestock an acoustic shield. Humans, wild canids, and domestic dogs therefore live in an evolutionary tangle of sound. Outside the forest, this intertwining is manifest in the sirens of emergency vehicles that call attention to themselves by wailing like über-wolves, tapping humanity's deep-rooted fears. Our domestic dogs hear the ancient echo also, howling at passing ambulances. The forest therefore follows us into civilization, buried in our psyches.

The howling stops as abruptly as it started. I am blind in the dark, and the coyotes' footfalls are silent, so I have no way of knowing whether or how the animals leave. Most likely they will slink away to their night's work, hunting small animals, guided by their own well-founded fears to circle widely around the human.

Silence returns to the mandala. I sink into the moment, feeling a familiar sense of arrival. The practice of returning to the mandala and sitting in silence for hundreds of hours has peeled back some of the barriers between the forest and my senses, intellect, and emotions. I can be present in a way that I had not known existed.

Despite this feeling of belonging, my relationship to this place is not straightforward. I simultaneously feel profound closeness and unutterable distance. As I have come to know the mandala, I have more clearly seen my ecological and evolutionary kinship with the forest. This knowledge feels woven into my body, remaking me or, more precisely, waking in me the ability to see how I was made all along.

At the same time, an equally powerful sense of otherness has grown. As I have watched, a realization of the enormity of my ignorance has pressed on me. Even simple enumeration and naming of the mandala's inhabitants lie far beyond my reach. An understanding of their lives and relationships in anything but a fragmentary way is quite impossible. The longer I watch, the more alienated I become from any hope of comprehending the mandala, of grasping its most basic nature.

Yet the separation that I feel is more than a heightened awareness of my ignorance. I have understood in some deep place that I am unnecessary here, as is all humanity. There is loneliness in this realization, poignancy in my irrelevance.

But I also feel an ineffable but strong sense of joy in the independence of the mandala's life. This was brought home to me several weeks ago as I walked into the forest. A hairy woodpecker lighted on a tree trunk and lobbed out its call. I was struck hard by the otherness of this bird. Here was a creature whose kind had chattered woodpecker calls for millions of years before humans came to be. Its daily existence was filled with bark flakes, hidden beetles, and the sounds of its woodpecker neighbors: another world, running parallel to my own. Millions of such parallel worlds exist in one mandala.

Somehow the shock of separateness flooded me with relief. The world does not center on me or on my species. The causal center of the natural world is a place that humans had no part in making. Life transcends us. It directs our gaze outward. I felt both humbled and uplifted by the woodpecker's flight.

So, I continue my watch, both stranger and kin at this mandala. The bright moon lifts the forest in a lambent, silver light. As my eyes adjust to the night, I see my shadow in the moonlight, resting across the circle of leaves.

Epilogue

It is commonplace for contemporary naturalists to deplore our culture's increasing disconnection from the natural world. I can sympathize with this complaint, at least in part. When asked to identify twenty corporate logos and twenty common species from our region, my first-year students can consistently name most of the corporate symbols and almost none of the species. The same would be true for most people in our culture.

But ours is not a new lament. Carl Linnaeus, one of the founders of modern ecology and taxonomy, wrote of the botanical abilities of his eighteenth-century compatriots, "few eyes see, and few minds understand. Through this want of observation and knowledge the world suffers immense loss." Much later, Aldo Leopold, reflecting on the state of the world in the 1940s, wrote, "Your true modern is separated from the land by many middlemen and physical gadgets. He has no vital relation to it . . . Turn him loose for a day on the land, and if the spot does not happen to be a golf links or a 'scenic area', he is bored stiff." It seems that skilled naturalists have always felt that their culture was perilously close to losing its last scrap of connection to the land.

Both men's words resonate with me, but I also feel that in some ways we now live in a better time for naturalists. Interest in the community of life is more widespread and vigorous than it has been for decades, perhaps centuries. Concern for the fate of ecosystems is part of our national and international political discourse. In less than a human lifetime, the fields of environmental activism, education, and science have grown from insignificance to prominence, and the question of how to heal our disconnection from nature has become a popular topic for educational reformers. All this interest is, perhaps, something new and encouraging. In Linnaeus's and Leopold's days, neither the popular imagination nor the government was much concerned with the ecology of other species. Of course, our modern interest is necessitated, in part, by the ecological mess that our forebears' insouciance bequeathed us, but I think it is also motivated by genuine interest in other forms of life and concern for their well-being.

Our modern world offers the naturalist many distractions and barriers, but it also provides a spectacular range of helpful tools. If Gilbert White, the eighteenth-century author of the classic *Natural History of Selborne*, had owned a library of accurate field guides, a computer with access to flower photographs and frog songs, and a database of the latest scientific papers, his close observations of nature could have been enriched, lessening his intellectual loneliness and giving him deeper ecological understanding. He could also, of course, have squandered his curiosity in synthetic worlds online, but the point here is that for those with an interest in natural history, we now have vastly more help available to us than at any other time.

It is with this help that I have explored the forest mandala. I hope this book will encourage others to start their own explorations. I was fortunate to be able to watch a small patch of old-growth forest. This is a rare privilege; old growth covers less than one-half of a percent of the land in the eastern United States. But old forests are not the only windows into the ecology of the world. Indeed, one outcome of my watch at the mandala has been to realize that we create wonderful places by giving them our attention, not by finding "pristine" places that will bring wonder to us. Gardens, urban trees, the sky, fields, young forests, a flock of suburban sparrows: these are all mandalas. Watching them closely is as fruitful as watching an ancient woodland.

We all differ in our ways of learning, so it is perhaps presumptuous of me to make suggestions for how to observe these mandalas. But two insights from my experience seem worth sharing with those who would like to try. The first is to leave behind expectations. Hoping for excitement, beauty, violence, enlightenment, or sacrament gets in the way of clear observation and will fog the mind with restlessness. Hope only for an enthusiastic openness of the senses.

The second suggestion is to borrow from the practice of meditation and to repeatedly return the mind's attention to the present moment. Our attention wanders, relentlessly. Bring it gently back. Over and over, seek out the sensory details: the particularities of sound, the feel and smell of the place, the visual complexities. This practice is not arduous, but it does take deliberate acts of the will.

The interior quality of our minds is itself a great teacher of natural history. It is here that we learn that "nature" is not a separate place. We too are animals, primates with a rich ecological and evolutionary context. By our paying attention, this inner animal can be watched at any time: our keen interest in fruits, meats, sugar, and salt; our obsession with social hierarchies, clans, and networks; our fascination with the aesthetics of human skin, hair, and bodily shapes; our incessant intellectual curiosity and ambition. Each one of us inhabits a storied mandala with as much complexity and depth as an old-growth forest. Even better, watching ourselves and watching the world are not in opposition; by observing the forest, I have come to see myself more clearly.

Part of what we discover by observing ourselves is an affinity for the world around us. The desire to name, understand, and enjoy the rest of the community of life is part of our humanity. Quiet observation of living mandalas offers one way to rediscover and develop this inheritance.

Acknowledgments

The mandala sits on land owned by the University of the South in Sewanee, Tennessee. Without the work of the many generations of people who have cared for this land, this book would not have been possible. My colleagues at the university provide a congenial and stimulating context in which to work. In particular, Nancy Berner, Jon Evans, Ann Fraser, John Fraser, Deborah McGrath, John Palisano, Jim Peters, Bran Potter, George Ramseur, Jean Yeatman, Harry Yeatman, and Kirk Zigler answered my questions about specific topics in this book. Jim Peters also gave me many insights into the nature of science, especially through our team-teaching of ecology and ethics. Conversations with Sid Brown and Tom Ward helped to put my experience of contemplative practice into a wider and more coherent context. DuPont Library's outstanding staff and excellent collections made research for this book a pleasure. The remarkable students at Sewanee give me inspiration and great hope for the future of biology and the study of natural history.

Walks in the forest with many local naturalists have also greatly expanded my appreciation of the natural history of our region. In particular, Joseph Bordley, Sanford McGee, and David Withers have shared many insights over the years.

Bill Hamilton, Stephen Kearsey, Beth Okamura, and Andrew Pomiankowski at the University of Oxford, and Chris Clark, Steve Emlen, Rick Harrison, Robert Johnston, Amy McCune, Carol McFadden,

Bobbi Peckarsky, Kern Reeve, Paul Sherman, and David Winkler at Cornell University were particularly generous and important mentors during my years of formal university training.

My fellow participants in the Wildbranch Writing Workshop at Sterling College helped me to grow as a writer and naturalist. I especially thank Tony Cross, Alison Hawthorne Deming, Jennifer Sahn, and Holly Wren Spaulding for their advice and example.

Early drafts of the manuscript benefited from editorial suggestions by John Gatta, Jean Haskell, George Haskell, and Jack Macrae. A modified version of the "Medicine" chapter was published by *Whole Terrain* and was improved by the work of Annie Jacobs and her editorial board. Henry Hamman was generous with his time, insights, and connections at a critical point in the development of the book.

Alice Martell is an extraordinary agent. Her perspicacious mentorship is a source of much encouragement and her splendid work brought this project to fruition. Kevin Doughten's insightful editorial direction brought coherence and vigor to the manuscript. His efforts as the book's shepherd, ambassador, and advocate have been outstanding.

I owe an immense intellectual debt to the thousands of naturalists whose scientific studies have deepened my understanding of biology. This book, I hope, honors their important work. My discussions necessarily omitted details from many of these studies, focusing on those parts that most directly touched my experience at the mandala or that helped me explain ideas in biology. This winnowing of detail is a dangerous business, especially in science, and so I encourage readers to dig into and beyond my bibliography to explore the richness of the topics that I have discussed here.

Sarah Vance supported this project with great generosity and insight. Her scientific critiques, editorial advice, and practical assistance with the preparation of the manuscript not only made the book possible but greatly increased its quality.

This book is a celebration of the life of forests, and so I will donate at least half my author's proceeds to projects that benefit forest conservation.

Bibliography

Preface

Bentley, G. E., ed. 2005. *William Blake: Selected Poems.* London: Penguin.

January 1st—Partnerships

Giles, H. A., trans. and ed. 1926. *Chuang Tzŭ.* 2nd ed., reprint 1980. London: Unwin Paperbacks.

Hale, M. E. 1983. *The Biology of Lichens.* 3rd ed. London: Edward Arnold.

Hanelt, B., and J. Janovy. 1999. "The life cycle of a horsehair worm, *Gordius robustus* (Nematomorpha: Gordioidea)." *Journal of Parasitology* 85: 139–41.

Hanelt, B., L. E. Grother, and J. Janovy. 2001. "Physid snails as sentinels of freshwater nematomorphs." *Journal of Parasitology* 87: 1049–53.

Nash, T. H., III, ed. 1996. *Lichen Biology.* Cambridge: Cambridge University Press.

Purvis, W. 2000. *Lichens.* Washington, DC: Smithsonian Institution Press.

Rivera, M. C., and J. A. Lake. 2004. "The ring of life provides evidence for a genome fusion origin of eukaryotes." *Nature* 431:152–55.

Thomas, F., A. Schmidt-Rhaesa, G. Martin, C. Manu, P. Durand, and F. Renaud. 2002. "Do hairworms (Nematomorpha) manipulate the water seeking behaviour of their terrestrial hosts?" *Journal of Evolutionary Biology* 15: 356–61.

January 17th—Kepler's Gift

Kepler, J. 1966. *The Six-Cornered Snowflake.* 1661. Translation and commentary by C. Hardie, B. J. Mason, and L. L. Whyte. Oxford: Clarendon Press.

Libbrecht, K. G. 1999. "A Snow Crystal Primer." Pasadena: California Institute of Technology. www.its.caltech.edu/~atomic/snowcrystals/primer/primer.htm.

Meinel, C. 1988. "Early seventeenth-century atomism: theory, epistemology, and the insufficiency of experiment." *Isis* 79: 68–103.

January 21st—The Experiment

Cimprich, D. A., and T. C. Grubb. 1994. "Consequences for Carolina Chickadees of foraging with Tufted Titmice in winter." *Ecology* 75: 1615–25.

Cooper, S. J., and D. L. Swanson. 1994. "Seasonal acclimatization of thermoregulation in the Black-capped Chickadee." *Condor* 96: 638–46.

Doherty, P. F., J. B. Williams, and T. C. Grubb. 2001. "Field metabolism and water flux of Carolina Chickadees during breeding and nonbreeding seasons: A test of the 'peak-demand' and 'reallocation' hypotheses." *Condor* 103: 370–75.

Gill, F. B. 2007. *Ornithology*. 3rd ed. New York: W. H. Freeman.

Grubb, T. C., Jr., and V. V. Pravasudov. 1994. "Tufted Titmouse (*Baeolophus bicolor*)," The Birds of North America Online (A. Poole, ed.). Ithaca, NY: Cornell Lab of Ornithology; doi:10.2173/bna.86.

Honkavaara, J., M. Koivula, E. Korpimäki, H. Siitari, and J. Viitala. 2002. "Ultraviolet vision and foraging in terrestrial vertebrates." *Oikos* 98: 505–11.

Karasov, W. H., M. C. Brittingham, and S. A. Temple. 1992. "Daily energy and expenditure by Black-capped Chickadees (*Parus atricapillus*) in winter." *Auk* 109: 393–95.

Marchand, P. J. 1991. *Life in the Cold*. 2nd ed. Hanover, NH: University Press of New England.

Mostrom, A. M., R. L. Curry, and B. Lohr. 2002. "Carolina Chickadee (*Poecile carolinensis*)." The Birds of North America Online. doi:10.2173/bna.636.

Norberg, R. A. 1978. "Energy content of some spiders and insects on branches of spruce (*Picea abies*) in winter: prey of certain passerine birds." *Oikos* 31: 222–29.

Pravosudov, V. V., T. C. Grubb, P. F. Doherty, C. L. Bronson, E. V. Pravosudova, and A. S. Dolby. 1999. "Social dominance and energy reserves in wintering woodland birds." *Condor* 101: 880–84.

Saarela, S., B. Klapper, and G. Heldmaier. 1995. "Daily rhythm of oxygen-consumption and thermoregulatory responses in some European winter-acclimatized or summer-acclimatized finches at different ambient-temperatures." *Journal of Comparative Physiology B: Biochemical, Systems, and Environmental Physiology* 165: 366–76.

Swanson, D. L., and E. T. Liknes. 2006. "A comparative analysis of thermogenic capacity and cold tolerance in small birds." *Journal of Experimental Biology* 209: 466–74.

Whittow, G. C., ed. 2000. *Sturkie's Avian Physiology*. 5th ed. San Diego: Academic Press.

January 30th—Winter Plants

Fenner, M., and K. Thompson. 2005. *The Ecology of Seeds*. Cambridge: Cambridge University Press.

Lambers, H., F. S. Chapin, and T. L. Pons. 1998. *Plant Physiological Ecology*. Berlin: Springer-Verlag.

Sakai, A., and W. Larcher. 1987. *Frost Survival of Plants: Responses and Adaptation to Freezing Stress*. Berlin: Springer-Verlag.

Taiz, L., and E. Zeiger. 2002. *Plant Physiology*. 3rd ed. Sunderland, MA: Sinauer Associates.

February 2nd—Footprints

Allen, J. A. 1877. *History of the American Bison.* Washington, DC: U.S. Department of the Interior.

Barlow, C. 2001. "Anachronistic fruits and the ghosts who haunt them." *Arnoldia* 61: 14–21.

Clarke, R. T. J., and T. Bauchop, eds. 1977. *Microbial Ecology of the Gut.* New York: Academic Press.

Delcourt, H. R., and P. A. Delcourt. 2000. "Eastern deciduous forests." In *North American Terrestrial Vegetation,* 2nd ed., edited by M. G. Barbour and W. D. Billings, 357–95. Cambridge: Cambridge University Press.

Gill, J. L., J. W. Williams, S. T. Jackson, K. B. Lininger, and G. S. Robinson. 2009. "Pleistocene megafaunal collapse, novel plant communities, and enhanced fire regimes in North America." *Science* 326: 1100–1103.

Graham, R. W. 2003. "Pleistocene tapir from Hill Top Cave, Trigg County, Kentucky, and a review of Plio-Pleistocene tapirs of North America and their paleoecology." In *Ice Age Cave Faunas of North America,* edited by B. W. Schubert, J. I. Mead, and R. W. Graham, 87–118. Bloomington: Indiana University Press.

Harriot, T. 1588. *A Briefe and True Report of the New Found Land of Virginia.* Reprint, 1972. New York: Dover Publications.

Hicks, D. J., and B. F. Chabot. 1985. "Deciduous forest." In *Physiological Ecology of North American Plant Communities,* edited by B. F. Chabot and H. A. Mooney, 257–77. New York: Chapman and Hall.

Hobson, P. N., ed. 1988. *The Rumen Microbial Ecosystem.* Barking, UK: Elsevier Science Publishers.

Lange, I. M. 2002. *Ice Age Mammals of North America: A Guide to the Big, the Hairy, and the Bizarre.* Missoula, MT: Mountain Press.

Martin, P. S., and R. G. Klein. 1984. *Quaternary Extinctions.* Tucson: University of Arizona Press.

McDonald, H. G. 2003. "Sloth remains from North American caves and associated karst features." In *Ice Age Cave Faunas of North America,* edited by B. W. Schubert, J. I. Mead, and R. W. Graham, 1–16. Bloomington: Indiana University Press.

Salley, A. S., ed. 1911. *Narratives of Early Carolina, 1650–1708.* New York: Scribner's Sons.

February 16th—Moss

Bateman, R. M., P. R. Crane, W. A. DiMichele, P. R. Kendrick, N. P. Rowe, T. Speck, and W. E. Stein. 1998. "Early evolution of land plants: phylogeny, physiology, and ecology of the primary terrestrial radiation." *Annual Review of Ecology and Systematics* 29: 263–92.

Conrad, H. S. 1956. *How to Know the Mosses and Liverworts.* Dubuque, IA: W. C. Brown.

Goffinet, B., and A. J. Shaw, eds. 2009. *Bryophyte Biology.* 2nd ed. Cambridge: Cambridge University Press.

Qiu, Y.-L., L. Li, B. Wang, Z. Chen, V. Knoop, M. Groth-Malonek, O. Dombrovska, J. Lee, L. Kent, J. Rest, G. F. Estabrook, T. A. Hendry, D. W. Taylor, C. M. Testa, M.

Ambros, B. Crandall-Stotler, R. J. Duff, M. Stech, W. Frey, D. Quandt, and C. C. Davis. 2006. "The deepest divergences in land plants inferred from phylogenomic evidence." *Proceedings of the National Academy of Sciences, USA* 103: 15511–16.

Qiu Y.-L., L. B. Li, B. Wang, Z. D. Chen, O. Dombrovska, J. J. Lee, L. Kent, R. Q. Li, R. W. Jobson, T. A. Hendry, D. W. Taylor, C. M. Testa, and M. Ambros. 2007. "A nonflowering land plant phylogeny inferred from nucleotide sequences of seven chloroplast, mitochondrial, and nuclear genes." *International Journal of Plant Sciences* 168: 691–708.

Richardson, D. H. S. 1981. *The Biology of Mosses*. New York: John Wiley and Sons.

February 28th—Salamander

Duellman, W. E., and L. Trueb. 1994. *Biology of Amphibians*. Baltimore: Johns Hopkins University Press.

Milanovich, J. R., W. E. Peterman, N. P. Nibbelink, and J. C. Maerz. 2010. "Projected loss of a salamander diversity hotspot as a consequence of projected global climate change." *PLoS ONE* 5: e12189. doi:10.1371/journal.pone.0012189.

Petranka, J. W. 1998. *Salamanders of the United States and Canada*. Washington, DC: Smithsonian Institution Press.

Petranka, J. W., M. E. Eldridge, and K. E. Haley. 1993. "Effects of timber harvesting on Southern Appalachian salamanders." *Conservation Biology* 7: 363–70.

Ruben, J. A., and A. J. Boucot. 1989. "The origin of the lungless salamanders (Amphibia: Plethodontidae)." *American Naturalist* 134: 161–69.

Stebbins, R. C., and N. W. Cohen. 1995. *A Natural History of Amphibians*. Princeton, NJ: Princeton University Press.

Vieites, D. R., M.-S. Min, and D. B. Wake. 2007. "Rapid diversification and dispersal during periods of global warming by plethodontid salamanders." *Proceedings of the National Academy of Sciences, USA* 104: 19903–7.

March 13th—*Hepatica*

Bennett, B. C. 2007. "Doctrine of Signatures: an explanation of medicinal plant discovery or dissemination of knowledge?" *Economic Botany* 61: 246–55.

Hartman, F. 1929. *The Life and Doctrine of Jacob Boehme*. New York: Macoy.

McGrew, R. E. 1985. *Encyclopedia of Medical History*. New York: McGraw-Hill.

March 13th—Snails

Chase, R. 2002. *Behavior and Its Neural Control in Gastropod Molluscs*. Oxford: Oxford University Press.

March 25th—Spring Ephemerals

Choe, J. C., and B. J. Crespi. 1997. *The Evolution of Social Behavior in Insects and Arachnids*. Cambridge: Cambridge University Press.

Curran, C. H. 1965. *The Families and Genera of North American* Diptera. Woodhaven, NY: Henry Tripp.

Motten, A. F. 1986. "Pollination ecology of the spring wildflower community of a temperate deciduous forest." *Ecological Monographs* 56: 21–42.

Sun, G., Q. Ji, D. L. Dilcher, S. Zheng, K. C. Nixon, and X. Wang. 2002. "Archaefructaceae, a new basal Angiosperm family." *Science* 296: 899–904.

Wilson, D. E., and S. Ruff. 1999. *The Smithsonian Book of North American Mammals*. Washington, DC: Smithsonian Institution Press.

April 2nd—Chainsaw

Duffy, D. C., and A. J Meier. 1992. "Do Appalachian herbaceous understories ever recover from clear-cutting?" *Conservation Biology* 6: 196–201.

Haskell, D. G., J. P. Evans, and N. W. Pelkey. 2006. "Depauperate avifauna in plantations compared to forests and exurban areas." *PLoS ONE* 1: e63. doi:10.1371/journal.pone.0000063.

Meier, A. J., S. P. Bratton, and D. C. Duffy. 1995. "Possible ecological mechanisms for loss of vernal-herb diversity in logged eastern deciduous forests." *Ecological Applications* 5: 935–46.

Perez-Garcia, J., B. Lippke, J. Comnick, and C. Manriquez. 2005. "An assessment of carbon pools, storage, and wood products market substitution using life-cycle analysis results." *Wood and Fiber Science* 37: 140–48.

Prestemon, J. P., and R. C. Abt. 2002. "Timber products supply and demand." Chap. 13 in *Southern Forest Resource Assessment*, edited by D. N. Wear and J. G. Greis. General Technical Report SRS-53, U.S. Department of Agriculture. Asheville, NC: Forest Service, Southern Research Station.

Scharai-Rad, M., and J. Welling. 2002. "Environmental and energy balances of wood products and substitutes." Rome: Food and Agriculture Organization of the United Nations. www.fao.org/docrep/004/y3609e/y3609e00.HTM.

Yarnell, S. 1998. *The Southern Appalachians: A History of the Landscape*. General Technical Report SRS-18, U.S. Department of Agriculture. Asheville, NC: Forest Service, Southern Research Station.

April 2nd—Flowers

Fenster, C. B., W. S Armbruster, P. Wilson, M. R. Dudash, and J. D. Thomson. 2004. "Pollination syndromes and floral specialization." *Annual Review of Ecology, Evolution, and Systematics* 35: 375–403.

Fosket, D. E. 1994. *Plant Growth and Development: A Molecular Approach*. San Diego: Academic Press.

Snow, A. A., and T. P. Spira. 1991. "Pollen vigor and the potential for sexual selection in plants." *Nature* 352: 796–97.

Walsh, N. E., and D. Charlesworth. 1992. "Evolutionary interpretations of differences in pollen-tube growth-rates." *Quarterly Review of Biology* 67: 19–37.

April 8th—Xylem

Ennos, R. 2001. *Trees*. Washington, DC: Smithsonian Institution Press.

Hacke, U. G., and J. S. Sperry. 2001. "Functional and ecological xylem anatomy." *Perspectives in Plant Ecology, Evolution and Systematics* 4: 97–115.

Sperry, J. S., J. R. Donnelly, and M. T. Tyree. 1988. "Seasonal occurrence of xylem embolism in sugar maple (*Acer saccharum*)." *American Journal of Botany* 75: 1212–18.

Tyree, M. T., and M. H. Zimmermann. 2002. *Xylem Structure and the Ascent of Sap*. 2nd ed. Berlin: Springer-Verlag.

April 14th—Moth

Smedley, S. R., and T. Eisner. 1996. "Sodium: a male moth's gift to its offspring." *Proceedings of the National Academy of Sciences, USA* 93: 809–13.

Young, M. 1997. *The Natural History of Moths*. London: T. and A. D. Poyser.

April 16th—Sunrise Birds

Pedrotti, F. L., L. S. Pedrotti, and L. M. Pedrotti. 2007. *Introduction to Optics*. 3rd ed. Upper Saddle River, NJ: Pearson Prentice Hall.

Wiley, R. H., and D. G. Richards. 1978. "Physical constraints on acoustic communication in the atmosphere: implications for the evolution of animal vocalizations." *Behavioral Ecology and Sociobiology* 3: 69–94.

April 22nd—Walking Seeds

Beattie, A., and D. C. Culver. 1981. "The guild of myrmecochores in a herbaceous flora of West Virginia forests." *Ecology* 62: 107–15.

Cain, M. L., H. Damman, and A. Muir. 1998. "Seed dispersal and the holocene migration of woodland herbs." *Ecological Monographs* 68: 325–47.

Clark, J. S. 1998. "Why trees migrate so fast: confronting theory with dispersal biology and the paleorecord." *American Naturalist* 152: 204–24.

Ness, J. H. 2004. "Forest edges and fire ants alter the seed shadow of an ant-dispersed plant." *Oecologia* 138: 448–54.

Smith, B. H., P. D. Forman, and A. E. Boyd. 1989. "Spatial patterns of seed dispersal and predation of two myrmecochorous forest herbs." *Ecology* 70: 1649–56.

Vellend, M., Myers, J. A., Gardescu, S., and P. L. Marks. 2003. "Dispersal of *Trillium* seeds by deer: implications for long-distance migration of forest herbs." *Ecology* 84: 1067–72.

April 29th—Earthquake

U.S. Geological Survey, Earthquake Hazards Program. "Magnitude 4.6 Alabama." http://neic.usgs.gov/neis/eq_depot/2003/eq_030429/.

May 7th—Wind

Ennos, A. R. 1997. "Wind as an ecological factor." *Trends in Ecology and Evolution* 12: 108–11.

Vogel, S. 1989. "Drag and reconfiguration of broad leaves in high winds." *Journal of Experimental Botany* 40: 941–48.

May 18th—Herbivory

Ananthakrishnan, T. N., and A. Raman. 1993. *Chemical Ecology of Phytophagous Insects.* New York: International Science Publisher.

Chown, S. L., and S. W. Nicolson. 2004. *Insect Physiological Ecology.* Oxford: Oxford University Press.

Hartley, S. E., and C. G. Jones. 2009. "Plant chemistry and herbivory, or why the world is green." In *Plant Ecology*, edited by M. J. Crawley. 2nd ed. Oxford: Blackwell Publishing.

Nation, J. L. 2008. *Insect Physiology and Biochemistry.* Boca Raton, FL: CRC Press.

Waldbauer, G. 1993. *What Good Are Bugs?: Insects in the Web of Life.* Cambridge, MA: Harvard University Press.

May 25th—Ripples

Clements, A. N. 1992. *The Biology of Mosquitoes: Development, Nutrition, and Reproduction.* London: Chapman and Hall.

Hames, R. S., K. V. Rosenberg, J. D. Lowe, S. E. Barker, and A. A. Dhondt. 2002. "Adverse effects of acid rain on the distribution of Wood Thrush *Hylocichla mustelina* in North America." *Proceedings of the National Academy of Sciences, USA* 99: 11235–40.

Spielman, A., and M. D'Antonio. 2001. *Mosquito: A Natural History of Our Most Persistent and Deadly Foe.* New York: Hyperion.

Whittow, G. C., ed. 2000. *Sturkie's Avian Physiology.* 5th ed. San Diego: Academic Press.

June 2nd—Quest

Klompen, H., and D. Grimaldi. 2001. "First Mesozoic record of a parasitiform mite: a larval Argasid tick in Cretaceous amber (Acari: Ixodida: Argasidae)." *Annals of the Entomological Society of America* 94: 10–15.

Sonenshine, D. E. 1991. *Biology of Ticks.* Oxford: Oxford University Press.

June 10th—Ferns

Schneider, H., E. Schuettpelz, K. M. Pryer, R. Cranfill, S. Magallon, and R. Lupia. 2004. "Ferns diversified in the shadow of angiosperms." *Nature* 428: 553–57.

Smith, A. R., K. M. Pryer, E. Schuettpelz, P. Korall, H. Schneider, and P. G. Wolf. 2006. "A classification for extant ferns." *Taxon* 55:705–31.

June 20th—A Tangle

Haase, M., and A. Karlsson. 2004. "Mate choice in a hermaphrodite: you won't score with a spermatophore." *Animal Behaviour* 67: 287–91.

Locher, R., and B. Baur. 2000. "Mating frequency and resource allocation to male and female function in the simultaneous hermaphrodite land snail *Arianta arbustorum*." *Journal of Evolutionary Biology* 13: 607–14.

Rogers, D. W., and R. Chase. 2002. "Determinants of paternity in the garden snail *Helix aspersa*." *Behavioral Ecology and Sociobiology* 52: 289–95.

Webster, J. P., J. I. Hoffman, and M. A. Berdoy. 2003. "Parasite infection, host resistance and mate choice: battle of the genders in a simultaneous hermaphrodite." *Proceedings of the Royal Society, Series B: Biological Sciences* 270: 1481–85.

July 2nd—Fungi

Hurst, L. D. 1996. "Why are there only two sexes?" *Proceedings of the Royal Society, Series B: Biological Sciences* 263: 415–22.

Webster, J., and R. W. S. Weber. 2007. *Introduction to Fungi*. 3rd ed. Cambridge: Cambridge University Press.

Whitfield, J. 2004. "Everything you always wanted to know about sexes." *PLoS Biol* 2(6): e183. doi:10.1371/journal.pbio.0020183.

Xu, J. 2005. "The inheritance of organelle genes and genomes: patterns and mechanisms." *Genome* 48: 951–58.

Yan, Z., and J. Xu. 2003. "Mitochondria are inherited from the MATa parent in crosses of the Basidiomycete fungus *Cryptococcus neoformans*." *Genetics* 163: 1315–25.

July 13th—Fireflies

Eisner, T., M. A. Goetz, D. E. Hill, S. R. Smedley, and J. Meinwald. 1997. "Firefly 'femmes fatales' acquire defensive steroids (lucibufagins) from their firefly prey." *Proceedings of the National Academy of Sciences, USA* 94: 9723–28.

July 27th—Sunfleck

Heinrich, B. 1996. *The Thermal Warriors: Strategies of Insect Survival*. Cambridge, MA: Harvard University Press.

Hull, J. C. 2002. "Photosynthetic induction dynamics to sunflecks of four deciduous forest understory herbs with different phenologies." *International Journal of Plant Sciences* 163: 913–24.

Williams, W. E., H. L. Gorton, and S. M. Witiak. 2003. "Chloroplast movements in the field." *Plant Cell and Environment:* 2005–14.

August 1st—Eft and Coyote

Brodie, E. D. 1968. "Investigations on the skin toxin of the Red-Spotted Newt, *Notophthalmus viridescens viridescens.*" *American Midland Naturalist* 80:276–80.

Hampton, B. 1997. *The Great American Wolf.* New York: Henry Holt and Company.

Parker, G. 1995. *Eastern Coyote: The Story of Its Success.* Halifax, Nova Scotia: Nimbus Publishing.

August 8th—Earthstar

Hibbett, D. S., E. M. Pine, E. Langer, G. Langer, and M. J. Donoghue. 1997. "Evolution of gilled mushrooms and puffballs inferred from ribosomal DNA sequences." *Proceedings of the National Academy of Sciences, USA* 94: 12002–6.

August 26th—Katydid

Capinera, J. L., R. D. Scott, and T. J. Walker. 2004. *Field Guide to Grasshoppers, Katydids, and Crickets of the United States.* Ithaca, NY: Cornell University Press.

Gerhardt, H. C., and F. Huber. 2002. *Acoustic Communication in Insects and Anurans.* Chicago: University of Chicago Press.

Gwynne, D. T. 2001. *Katydids and Bush-Crickets: Reproductive Behavior and Evolution of the Tettigoniidae.* Ithaca, NY: Cornell University Press.

Rannels, S., W. Hershberger, and J. Dillon. 1998. *Songs of Crickets and Katydids of the Mid-Atlantic States.* CD audio recording. Maugansville, MD: Wil Hershberger.

September 21st—Medicine

Culpeper, N. 1653. *Culpeper's Complete Herbal.* Reprint, 1985. Secaucus, NJ: Chartwell Books.

Horn, D., T. Cathcart, T. E. Hemmerly, and D. Duhl, eds. 2005. *Wildflowers of Tennessee, the Ohio Valley, and the Southern Appalachians.* Auburn, WA: Lone Pine Publishing.

Lewis, W. H., and M. P. F. Elvin-Lewis. 1977. *Medical Botany: Plants Affecting Man's Health.* New York: John Wiley and Sons.

Mann, R. D. 1985. *William Withering and the Foxglove.* Lancaster, UK: MTP Press.

Moerman, D. E. 1998. *Native American Ethnobotany.* Portland, OR: Timber Press.

U.S. Fish and Wildlife Service. 2009. *General Advice for the Export of Wild and Wild-Simulated American Ginseng* (Panax quinquefolius) *Harvested in 2009 and 2010 from States with Approved CITES Export Programs.* Washington, DC: U.S. Department of the Interior.

Vanisree, M., C.-Y. Lee, S.-F. Lo, S. M. Nalawade, C. Y. Lin, and H.-S. Tsay. 2004. "Studies on the production of some important secondary metabolites from medicinal plants by plant tissue cultures." *Botanical Bulletin of Academia Sinica* 45: 1–22.

September 23rd—Caterpillar

Heinrich, B. 2009. *Summer World: A Season of Bounty*. New York: Ecco.

Heinrich, B., and S. L. Collins. 1983. "Caterpillar leaf damage, and the game of hide-and-seek with birds." *Ecology* 64: 592–602.

Real, P. G., R. Iannazzi, A. C. Kamil, and B. Heinrich. 1984. "Discrimination and generalization of leaf damage by blue jays (*Cyanocitta cristata*)." *Animal Learning and Behavior* 12: 202–8.

Stamp, N. E., and T. M. Casey, eds. 1993. *Caterpillars: Ecological and Evolutionary Constraints on Foraging*. London: Chapman and Hall.

Wagner, D. L. 2005. *Caterpillars of Eastern North America: A Guide to Identification and Natural History*. Princeton, NJ: Princeton University Press.

September 23rd—Vulture

Blount, J. D., D. C. Houston, A. P. Møller, and J. Wright. 2003. "Do individual branches of immune defence correlate? A comparative case study of scavenging and non-scavenging birds." *Oikos* 102: 340–50.

DeVault, T. L., O. E. Rhodes, Jr., and J. A. Shivik. 2003. "Scavenging by vertebrates: behavioral, ecological, and evolutionary perspectives on an important energy transfer pathway in terrestrial ecosystems." *Oikos* 102:225–34.

Kelly, N. E., D. W. Sparks, T. L. DeVault, and O. E. Rhodes, Jr. 2007. "Diet of Black and Turkey Vultures in a forested landscape." *Wilson Journal of Ornithology* 119: 267–70.

Kirk, D. A., and M. J. Mossman. 1998. "Turkey Vulture (*Cathartes aura*)," The Birds of North America Online (A. Poole, ed.). Ithaca, NY: Cornell Lab of Ornithology. doi:10.2173/bna.339.

Markandya, A., T. Taylor, A. Longo, M. N. Murty, S. Murty, and K. Dhavala. 2008. "Counting the cost of vulture decline—An appraisal of the human health and other benefits of vultures in India." *Ecological Economics* 67: 194–204.

Powers, W. *The Science of Smell*. Iowa State University Extension. www.extension.iastate.edu/Publications/PM1963a.pdf.

September 26th—Migrants

Evans Ogden, L. J., and B. J. Stutchbury. 1994. "Hooded Warbler (*Wilsonia citrina*)," The Birds of North America Online (A. Poole, ed.). Ithaca, NY: Cornell Lab of Ornithology. doi:10.2173/bna.110.

Hughes, J. M. 1999. "Yellow-billed Cuckoo (*Coccyzus americanus*)," The Birds of North America Online (A. Poole, ed.). Ithaca, NY: Cornell Lab of Ornithology. doi:10.2173/bna.418.

Rimmer, C. C., and K. P. McFarland. 1998. "Tennessee Warbler (*Vermivora peregrina*)," The Birds of North America Online. doi:10.2173/bna.350.

October 5th—Alarm Waves

Agrawal, A. A. 2000. "Communication between plants: this time it's real." *Trends in Ecology and Evolution* 15: 446.

Caro, T. M., L. Lombardo, A. W. Goldizen, and M. Kelly. 1995. "Tail-flagging and other antipredator signals in white-tailed deer: new data and synthesis." *Behavioral Ecology* 6: 442–50.

Cotton, S. 2001. "Methyl jasmonate." www.chm.bris.ac.uk/motm/jasmine/jasminev.htm.

Farmer, E. E., and C. A. Ryan. 1990. "Interplant communication: airborne methyl jasmonate induces synthesis of proteinase inhibitors in plant leaves." *Proceedings of the National Academy of Sciences, USA* 87: 7713–16.

FitzGibbon, C. D., and J. H. Fanshawe. 1988. "Stotting in Thomson's gazelles: an honest signal of condition." *Behavioral Ecology and Sociobiology* 23: 69–74.

Maloof, J. 2006. "Breathe." *Conservation in Practice* 7: 5–6.

October 14th—Samara

Green, D. S. 1980. "The terminal velocity and dispersal of spinning samaras." *American Journal of Botany* 67: 1218–24.

Horn, H. S., R. Nathan, and S. R. Kaplan. 2001. "Long-distance dispersal of tree seeds by wind." *Ecological Research* 16: 877–85.

Lentink, D., W. B. Dickson, J. L. van Leewen, and M. H. Dickinson. 2009. "Leading-edge vortices elevate lift of autorotating plant seeds." *Science* 324: 1438–40.

Sipe, T. W., and A. R. Linnerooth. 1995. "Intraspecific variation in samara morphology and flight behavior in *Acer saccharinum* (Aceraceae)." *American Journal of Botany* 82: 1412–19.

October 29th—Faces

Darwin, C. 1872. *The Expression of the Emotions in Man and Animals*. Reprint, 1965. Chicago: University of Chicago Press.

Lorenz, K. 1971. *Studies in Animal and Human Behaviour.* Translated by R. Martin. Cambridge, MA: Harvard University Press.

Randall, J. A. 2001. "Evolution and function of drumming as communication in mammals." *American Zoologist* 41: 1143–56.

Todorov, A., C. P. Said, A. D. Engell, and N. N. Oosterhof. 2008. "Understanding evaluation of faces on social dimensions." *Trends in Cognitive Sciences* 12: 455–60.

November 5th—Light

Caine, N. G., D. Osorio, and N. I. Mundy. 2009. "A foraging advantage for dichromatic marmosets (*Callithrix geoffroyi*) at low light intensity." *Biology Letters* 6: 36–38.

Craig, C. L., R. S. Weber, and G. D. Bernard. 1996. "Evolution of predator-prey systems: Spider foraging plasticity in response to the visual ecology of prey." *American Naturalist* 147: 205–29.

Endler, J. A. 2006. "Disruptive and cryptic coloration." *Proceedings of the Royal Society, Series B: Biological Sciences* 273: 2425–26.

———. 1997. "Light, behavior, and conservation of forest dwelling organisms." In *Behavioral Approaches to Conservation in the Wild*, edited by J. R. Clemmons and R. Buchholz, 329–55. Cambridge: Cambridge University Press.

King, R. B., S. Hauff, and J. B. Phillips. 1994. "Physiological color change in the green treefrog: Responses to background brightness and temperature." *Copeia* 1994: 422–32.

Merilaita, S., and J. Lind. 2005. "Background-matching and disruptive coloration, and the evolution of cryptic coloration." *Proceedings of the Royal Society, Series B: Biological Sciences* 272: 665–70.

Mollon, J. D., J. K. Bowmaker, and G. H. Jacobs. 1984. "Variations of color-vision in a New World primate can be explained by polymorphism of retinal photopigments." *Proceedings of the Royal Society, Series B: Biological Sciences* 222: 373–99.

Morgan, M. J., A. Adam, and J. D. Mollon. 1992. "Dichromats detect colour-camouflaged objects that are not detected by trichromats." *Proceedings of the Royal Society, Series B: Biological Sciences* 248: 291–95.

Schaefer, H. M., and N. Stobbe. 2006. "Disruptive coloration provides camouflage independent of background matching." *Proceedings of the Royal Society, Series B: Biological Sciences* 273: 2427–32.

Stevens, M., I. C. Cuthill, A. M. M. Windsor, and H. J. Walker. 2006. "Disruptive contrast in animal camouflage." *Proceedings of the Royal Society, Series B: Biological Sciences* 273: 2433–38.

November 15th—Sharp-shinned Hawk

Bildstein, K. L., and K. Meyer. 2000. "Sharp-shinned Hawk (*Accipiter striatus*)," The Birds of North America Online (A. Poole, ed.). Ithaca, NY: Cornell Lab of Ornithology. doi:10.2173/bna.482.

Hughes, N. M., H. S. Neufeld, and K. O. Burkey. 2005. "Functional role of anthocyanins in high-light winter leaves of the evergreen herb *Galax urceolata*." *New Phytologist* 168: 575–87.

Lin, E. 2005. *Production and Processing of Small Seeds for Birds*. Agricultural and Food Engineering Technical Report 1. Rome: Food and Agriculture Organization of the United Nations.

Marden, J. H. 1987. "Maximum lift production during takeoff in flying animals." *Journal of Experimental Biology* 130: 235–38.

Zhang, J., G. Harbottle, C. Wang, and Z. Kong. 1999. "Oldest playable musical instruments found at Jiahu early Neolithic site in China." *Nature* 401: 366–68.

November 21st—Twigs

Canadell, J. G., C. Le Quere, M. R. Raupach, C. B. Field, E. T. Buitenhuis, P. Ciais, T. J. Conway, N. P. Gillett, R. A. Houghton, and G. Marland. 2007. "Contributions to accelerating atmospheric CO_2 growth from economic activity, carbon intensity, and efficiency of natural sinks." *Proceedings of the National Academy of Sciences, USA* 104: 18866–70.

Dixon R. K., A. M. Solomon, S. Brown, R. A. Houghton, M. C. Trexier, and J. Wisniewski. 1994. "Carbon pools and flux of global forest ecosystems." *Science* 263: 185–90.

Hopkins, W. G. 1999. *Introduction to Plant Physiology*. 2nd ed. New York: John Wiley and Sons.

Howard, J. L. 2004. *Ailanthus altissima*. In: Fire Effects Information System. U.S. Department of Agriculture, Forest Service, Rocky Mountain Research Station. www.fs.fed.us/database/feis/plants/tree/ailalt/all.html.

Innes, R. J. 2009. *Paulownia tomentosa*. In: Fire Effects Information System. www.fs.fed.us/database/feis/plants/tree/pautom/all.html.

Solomon, S., D. Qin, M. Manning, Z. Chen, M. Marquis, K. B. Averyt, M. Tignor, and H. L. Miller (eds.). 2007. *Contribution of Working Group I to the Fourth Assessment Report of the Intergovernmental Panel on Climate Change*. Cambridge: Cambridge University Press.

Woodbury, P. B., J. E. Smith, and L. S. Heath 2007. "Carbon sequestration in the U.S. forest sector from 1990 to 2010." *Forest Ecology and Management* 241: 14–27.

December 3rd—Litter

Coleman, D. C., and D. A. Crossley, Jr. 1996. *Fundamentals of Soil Ecology*. San Diego: Academic Press.

Crawford, J. W., J. A. Harris, K. Ritz, and I. M. Young. 2005. "Towards an evolutionary ecology of life in soil." *Trends in Ecology and Evolution* 20: 81–87.

Horton, T. R., and T. D. Bruns. 2001. "The molecular revolution in ectomycorrhizal ecology: peeking into the black-box." *Molecular Ecology* 10: 1855–71.

Wolfe, D. W. 2001. *Tales from the Underground: A Natural History of Subterranean Life*. Reading, MA: Perseus Publishing.

December 6th—Underground Bestiary

Budd, G. E., and M. J. Telford. 2009. "The origin and evolution of arthropods." *Nature* 457: 812–17.

Hopkin, S. P. 1997. *Biology of the Springtails (Insecta: Collembola)*. Oxford: Oxford University Press.

Regier, J. C., J. W. Shultz, A. Zwick, A. Hussey, B. Ball, R. Wetzer, J. W. Martin, and C. W. Cunningham. 2010. "Arthropod relationships revealed by phylogenomic analysis of nuclear protein-coding sequences." *Nature* 463: 1079–83.

Ruppert, E. E., R. S. Fox, and R. D. Barnes. 2004. *Invertebrate Zoology: A Functional Evolutionary Approach*. 7th ed. Belmont, CA: Brooks/Cole-Thomson Learning.

December 26th—Treetops

Weiss, R. 2003. "Administration opens Alaska's Tongass forest to logging." *The Washington Post*, December 24, page A16.

December 31st—Watching

Bender, D. J., E. M. Bayne, and R. M. Brigham. 1996. "Lunar condition influences coyote (*Canis latrans*) howling." *American Midland Naturalist* 136: 413–17.

Gese, E. M., and R. L. Ruff. 1998. "Howling by coyotes (*Canis latrans*): variation among social classes, seasons, and pack sizes." *Canadian Journal of Zoology* 76: 1037–43.

Epilogue

Davis, M. B., ed. 1996. *Eastern Old-Growth Forest: Prospects for Rediscovery and Recovery.* Washington, DC: Island Press.

Leopold, A. 1949. *A Sand County Almanac, and Sketches Here and There.* New York: Oxford University Press.

Linnaeus, C. [1707–1788], quoted as epigram in Nicholas Culpeper, *The English Physician*, edited by E. Sibly. Reprint, 1800. London: Satcherd.

White, G. 1788–89. *The Natural History of Selbourne*, edited by R. Mabey. Reprint, 1977. London: Penguin Books.

Index